STUDENT SOLUTIONS MANUAL FOR

PHYSICS
Algebra/Trig

EUGENE HECHT
Adelphi University

Brooks/Cole Publishing Company
Pacific Grove, California

I(T)P® **An International Thomson Publishing Company**

Brooks/Cole Publishing Company
A Division of International Thomson Publishing Inc.

Printed in the United States of America

10 9 8 7 6 5 4 3

ISBN 0-534-09118-0

Sponsoring Editor: *Audra C. Silverie*
Editorial Assistant: *Beth Wilbur*
Production Coordinator: *Dorothy Bell*
Cover Design: *E. Kelly Shoemaker*
Typesetting: *Laurel Technical Services*
Printing and Binding: *Malloy Lithographing, Inc.*

PREFACE

This Student Solutions Manual was written to complement *Physics* by Eugene Hecht. It is designed to assist you in working independently to master the material discussed in the text.

Every chapter in the main text contains approximately twenty Discussion Questions. A selection of the odd-numbered questions are listed with italic numerals, and the answers to those questions appear in this manual. The Discussion Questions are meant to help you explore and develop your conceptual comprehension of the material. If an answer is not apparent, reread the appropriate portions of the text. The answers provided in this manual should be consulted only as a last resort or, alternatively, as a check.

Every chapter in the main text also contains approximately twenty Multiple Choice Questions. These are designed to test and develop your ability to deal quickly and accurately with conceptual issues that require little or no mathematical processing. Read the questions carefully. Make no assumptions about what the author might have had in mind. Respond only to the question as stated. This manual contains answers to the odd-numbered Multiple Choice Questions which are also answered in the main text.

Every chapter in the main text also contains about eighty numerical Problems listed according to increasing difficulty. A representative selection of the odd-numbered problems have italic numerals. Complete solutions for all of these problems appear in this manual. Again you are advised to attempt to solve each problem before consulting the Student Solutions Manual. The Answer Section of the main text contains a selection of skeletal solutions. Study the solutions for similar problems before turning to the complete solutions provided in this manual.

Many people contributed to this project and deserve recognition, particularly Audra Silverie, Assistant Editor; Beth Wilbur, Editorial Associate; and Lisa Moller, Editor. Also, a special thanks to the staff of Laurel Technical Services for their patience and understanding, especially Kurt Norlin and Terri Bittner.

Suggestions, corrections, and criticisms should be sent to Professor Gene Hecht, Physics Dept., Adelphi University, Garden City, N.Y. 11530. Good Luck!

CONTENTS

Answers to Discussion Questions

1.1   Because on rare occasions there have been scientists who
        have simply lied about their work. The biologist who, in his
        laboratory, colored the butt of a white mouse with a black
        magic marker and then claimed to have performed a skin graft
        wasn't doing science. There are also a few cases of honest
        self-delusion in which what scientists have done has been
        utter nonsense and certainly not science.

1.3   Laws transcend experience because they generalize. We know
        that something has happened the same way for years and from
        that we conclude that it will continue to happen that way.
        Laws are timeless, and that goes beyond experience. The same
        can be said about the application of law everywhere — it
        would be strange if certain physical laws only worked in
        Chicago. Still, we certainly don't test any of them
        everywhere. Whenever we say that something happens all the
        time or everywhere, we are making an inference based on a
        finite number of observations. We have to work this way,
        since it would be impossible to test the applicability of a
        proposed law in all possible instances before endorsing it.

1.5   Science seeks to understand the Universe as we perceive it
        to be. This is to be done as best we can, knowing the
        tentative nature of theory. At the same time, we must be
        economical in our formulation. As William of Ockham
        (ca. 1290-1349) suggested: the simplest, briefest, least
        complex explanation is the one to be accepted.

1.7   Data are the uninterpreted perceptions (as much as that's
        possible). Facts are the product of interpretation within
        the context of some world view, some theoretical
        understanding. "I see a disc of light there" becomes "I see
        the sun god in heaven" or "I see a star in space." These are
        statements of fact based on different world views. Since
        world views can be biased, facts can be wrong.

1.9   The ultimate testing ground in physics is nature itself. If
        the universe is found experimentally to match the
        predictions of the theory, then it was "crazy enough."

Answers to Multiple Choice Questions

1. a   3. a   5. c   7. d   9. d   11. d   13. d

15. b   17. a   19. c

Solutions to Problems

1.1   Ten billion = $10\,000\,000\,000 = 10^{10}$.

1.9   Since there are ten Ångstroms in one nanometer,
$$5 \times 10^3 \text{ Å} = \frac{5 \times 10^3}{10} \text{ nm} = 5 \times 10^2 \text{ nm}.$$

1.19   Since there are 2.54 cm in one inch,
$$65 \text{ cm}^2 = \frac{65}{2.54^2} \text{ in.}^2 = 10 \text{ in.}^2.$$

So the dog's sensory area is
$(10 \text{ in.}^2)/(3/4 \text{ in.}^2) = 13.3 = 13$ times greater than the human's.

1.23   For one significant figure, we take the *first* digit from the left, 2, and round up because the next digit, 9, is greater than 5. Then we convert to scientific notation: 3×10^8 m/s.

For three significant figures, we perform the same operation starting with the *third* digit from the left.
Result: 3.00×10^8 m/s.

For four significant figures, we get 2.998×10^8 m/s, and for eight significant figures, $2.997\,924\,6 \times 10^8$ m/s.

1.31   Since there are 2.540 cm in an inch, there are
$$\frac{1}{2.540} = 0.393\,7 \text{ inches in a cm.}$$

Conversion would consist of multiplying by 0.393 7.

1.39   (6 × 10⁵ particles/h)(24 h/d)(365 ¼ d/y)
       = 5.259 6 × 10⁹ particles per year.

       Their total mass is (1.5 lb)(0.453 6 kg/lb) = 0.680 4 kg.

       The mass of one particle is, then,
       (0.680 4 kg)/(5.259 6 × 10⁹) = 1.293 6 × 10⁻¹⁰ kg = 1 × 10⁻⁷ g.

       In 50 years, the total mass will be
                 50(0.68 kg) = 34 kg = 0.3 × 10² kg.

1.43   With all masses converted to grass, the sum is
       1.00 g + 0.001 00 g + 1.00 × 10³ g + 0.000 001 00 g
       = 1001.001 001 g = 1.00 kg.

Answers to Discussion Questions

2.1   If the average speed is zero, the object didn't move, so it
        is not possible for it to have a nonzero speed over a
        smaller time interval. The average velocity may well be zero
        even for a moving body if it returned to its original
        position. Hence, it could have a nonzero value over a
        shorter time interval.

2.5   Yes. Imagine two people starting out at 9:00 a.m. on the
        same day, one at the top going down, the other at the bottom
        going up. They must meet somewhere, and when they do, they
        will be in the same place at the same time.

2.7   Since speed is the magnitude of the velocity vector, a
        constant velocity means a constant speed. However, a vector
        can have constant magnitude while changing direction. So a
        constant speed does not necessarily mean a constant
        velocity. Constant speed without constant velocity is
        possible. A constant speed can occur in a changing
        direction.

2.9   The dog was already running when the clock started at $t = 0$,
        and he continued at a constant speed until $t = 2$ s, at which
        point he stopped, having gone 3 m. He rested until $t = 4$ s
        and then ran at a constant speed until $t = 10$ s, whereupon
        he put on a burst of speed. Nothing can be said about the
        displacement — the dog could be running in circles for all
        we know. He traveled a total distance of 6 m. He moved the
        fastest from 10 s to 11 s.

2.11   Yes, changing the units will change the numerical value of
        the magnitude of a vector. Its orientation, or direction in
        space, however, will not change.

2.13   The mouse began moving at $t = 1$ s, from a point 1 m in from
        the origin. It ran to a point 2 m from the origin in 1 s and
        stopped. It rested for 1 s and then, at $t = 2$ s, it started
        running at nonuniform speed until it got 6 m from the origin
        at $t = 6$ s. There, it immediately turned around and ran at a
        constant speed back to the opening of the tunnel in 2 s.

2.17 (a) At $t = 0$, the displacement is $0/DC = 0$.

   (b) For $t \gg C$, $s \approx (At^2 + Bt)/Dt$.

   (c) For $t \ll C$, $s \approx (At^2 + Bt)/DC$.

Answers to Multiple Choice Questions

1. d   3. b   5. c   7. c   9. c   11. d   13. b

15. a   17. c

Solutions to Problems

2.3   Speed $= \dfrac{\text{distance}}{\text{time}}$; symbolically, $v = 1/t$.

   (a)   For a distance of 1.0 ft in vacuum, the time is given
         by $t = 1/v = (1.0 \text{ ft})(0.304\ 8 \text{ m/ft})/(2.998 \times 10^8 \text{ m/s})$
         $= 1.0 \times 10^{-9}$ s.

   (b)   For a distance of 1000 m, the time is given by
         $t = 1/v = (1000 \text{ m})/(2.998 \times 10^8 \text{ m/s}) = 3.336 \times 10^{-6}$ s.

2.11 $\Delta t = 2.83 \text{ s} - 1.33 \text{ s} = 1.50$ s.

   Distance = speed·time; symbolically, $l = vt$.

   For the bee, traveling at a constant speed of
   10 m/s for 1.5 s, $l = (10 \text{ m/s})(1.50 \text{ s}) = 15$ m.

2.23 $s = \sqrt{(1.0 \text{ m})^2 + (1.0 \text{ m})^2} = 1.4$ m (color and body temperature
   are irrelevant).

2.37   The distance is given by $s = \sqrt{(20 \text{ m})^2 + (60 \text{ m})^2} = 63$ m.
   The angle of the displacement vector is determined by
   $\tan \theta = 60/20$. Then $\theta = 71.6°$. So **s** is a vector of magnitude
   63 m elevated 72° up from the horizontal.

2.49   Since the runners meet after 100 s, the fly is in the air for 100 s. At 10 m/s it travels 1000 m, or to two significant figures, 10×10^2 m.

2.59   (a)   The horizontal component is given by
$v_H = (300 \text{ m/s}) \cos 32.0° = 254$ m/s.

        (b)   The vertical component is given by
$v_V = (300 \text{ m/s}) \sin 32.0° = 159$ m/s.

        (c) $v = \sqrt{v_H^2 + v_V^2}$, so $\sqrt{254^2 + 159^2}$ m/s = $\sqrt{89\,797}$ m/s. It should, and does, approximately equal 300 m/s.

2.63   In 5.0 s the mouse moves 10 m north, hence the hypotenuse of the triangle is $\sqrt{(50 \text{ m})^2 + (10 \text{ m})^2}$ = 50.99 m.

The hawk's speed, then, must be (50.99 m)/(5.0 s) = 10.198 m/s or 10 m/s.

The dive angle, θ is given by
$\tan \theta = (10 \text{ m})/(50 \text{ m})$. $\theta = 11°$.

2.67   Let $\mathbf{v}_{SE}$ = the ship's velocity with respect to the earth,
        $\mathbf{v}_{JS}$ = the jogger's velocity with respect to the ship,
and $\mathbf{v}_{JE}$ = the jogger's velocity with respect to the earth.

$\mathbf{v}_{JE} = \mathbf{v}_{JS} + \mathbf{v}_{SE}$, and since $\mathbf{v}_{SE}$ and $\mathbf{v}_{JS}$ are perpendicular, v_{JE} is $\sqrt{v_{JS}^2 + v_{SE}^2} = \sqrt{(10 \text{ km/h})^2 + (20 \text{ km/h})^2}$ = 22 km/h.

The angle the jogger makes measured from the direction of the ship is given by $\sin \theta = (10 \text{ km/h})/(22 \text{ km/h})$.

So, $\theta = 27°$ which puts the direction of $\mathbf{v}_{SE}$ 3° west of north.

Answers to Discussion Questions

3.3   The balls are identical and separately must fall together, that is, at the same rate. When two are stuck together and dropped alongside a single ball, they again must fall together provided the air resistance on the double mass is no different. In vacuum, there is no air resistance, so both chunks of clay (one twice the mass of the other) must fall together. Indeed, 1000 such balls of clay stuck together and shaped into a piano would fall the same way — all things must fall together in vacuum.

3.7   The cars are always separated by one second. The distance each travels is proportional to the time squared, so their distance separation will increase: after 1 s, the separation will be $\frac{1}{2}a(1\ s)^2$; after 10 s, it will be $\frac{1}{2}a[(10\ s)^2 - (9\ s)^2] = \frac{1}{2}a(19\ s^2)$. The cars will be closest to each other at the starting line. The time between their successive crossings of the finish line will be 1 s.

3.9   For constant acceleration, $v_{av} = \frac{1}{2}(v_f + v_i)$, and the average is midway between the initial and final speeds. But with a increasing, a burst of speed at the end of the flight means that more time is spent at speeds closer to the initial speed. The average shifts down closer to the initial speed.

3.11   The keys fall at the same rate as the floor and so cannot get any closer to it. The keys float in front of your face, which is also falling at the same rate.

3.15 $a = [(At^2 - Bt)/(t + C)D] + Et^2/(t - C)D$.

(a)   When $t = 0$, $a = 0/CD + 0/(-CD) = 0$.

(b)   When $t \gg C$, $a \approx [(At^2 - Bt)/tD] + Et^2/tD$
     $= (At^2 - Bt + Et^2)/tD$.

(c)   When $t \ll C$, $a \approx [(At^2 - Bt)/CD] - Et^2/CD$
     $= (At^2 - Bt - Et^2)/CD$.

3.17   The speed reaches a maximum of 25 m/s at about 4.2 s, and then drops to zero at the peak altitude ($t_p \approx 7.4$ s). The net acceleration increases for about 2 s, after which it decreases as the engine throttles down. At $t = 4.2$ s, with the engine still firing, the net acceleration is zero as the engine's thrust and gravity cancel each other. Thereafter it is increasingly negative as the engine's thrust decreases and gravity dominates. The engine shuts off at around 5.5 s and the rocket decelerates while still moving upward, until it stops at $t \approx 7.4$ s.

Answers to Multiple Choice Questions

1. a   3. c   5. b   7. b   9. e   11. a   13. d

15. c   17. d

Solutions to Problems

3.1 $v_f = 100$ m/s, $v_i = 0$. The average acceleration is given by $a_{av} = (v_f - v_i)/\Delta t = (100 \text{ m/s})(10 \text{ s}) = 10 \text{ m/s}^2$.

3.5   The time elapsed is 62 s, so the average acceleration is given by $a_{av} = (v_f - v_i)/\Delta t = (15.0 \text{ m/s} - 1.0 \text{ m/s})/(62 \text{ s})$ $= 0.23 \text{ m/s}^2$.

3.9   From Fig. (3.2a), $a_{av} = \Delta v/\Delta t = \text{constant} = (25 \text{ m/s})/(5.0 \text{ s}) = 5.0 \text{ m/s}^2$.

3.31 $v_i = 2.2$ m/s, $v_f = 0$, $s = 10$ m. First we find a, which is constant.

$v_f^2 = v_i^2 + 2as$, so $a = -v_i^2/2s = -(2.2 \text{ m/s})^2/[2(10 \text{ m})]$ $= -0.242 \text{ m/s}^2$.

At time t, then, the distance traveled is $s = v_i t + \frac{1}{2}at^2$ $= (2.2 \text{ m/s})t + (-0.121 \text{ m/s}^2)t^2$.

$s_2 = v_i t + \frac{1}{2}at^2 = (2.2 \text{ m/s})(2.0 \text{ s}) +$ $\frac{1}{2}(-0.242 \text{ m/s}^2)(2.0 \text{ s})^2 = 3.92 \text{ m}$.

$s_3 = (2.2 \text{ m/s})(3.0 \text{ s}) +$
$\quad ½(-0.242 \text{ m/s}^2)(3.0 \text{ s})^2 = 5.51 \text{ m}. \; s_3 - s_2 = 1.6 \text{ m}.$

Since $s_3 - s_2 = 1.595 \text{ m} = 1.6 \text{ m}$, the swimmer travels about 1.6 m during the third second.

3.35 $s = v_i t + ½at^2$.

(a) In 1 s the first missile travels $s_1 = (60 \text{ m/s})(1 \text{ s})$
    $+ ½(20 \text{ m/s}^2)(1 \text{ s})^2 = 70 \text{ m}$.

(b) Two seconds after the first launch, the second missile has been in the air 1 s and so has traveled 70 m. The first missile, meanwhile, has traveled $(60 \text{ m/s})(2 \text{ s}) + ½(20 \text{ m/s}^2)(2 \text{ s})^2 = 160 \text{ m}$. The difference, 90 m, is the separation of the first and second missiles.

3.45 $v_i = 0$, so
        $v_f^2 = v_i^2 + 2gs = 0 + 2(9.81 \text{ m/s}^2)(0.50 \text{ m}) = 9.81 \text{ m}^2/\text{s}^2$.

Then $v_f = 3.132 \text{ m/s}$, or to two figures, 3.1 m/s.

Multiplying by 3.28 ft/m yields $v_f = 10 \text{ ft/s}$, and multiplying that by $\dfrac{1}{5280}$ mi/ft and by 3600 s/h yields $v_f = 6.9 \text{ mi/h}$.

3.67   A launch angle of less than 45° will cause the shell to return to launch elevation at a shorter distance from the launch point, but the flatter trajectory makes for more distance overall.

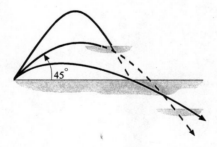

3.73 Let v_i be the initial speed, v_v the vertical speed at time t, and v_H the horizontal speed at time t. Then $v_v = v_i \sin \theta + gt$, $v_H = v_i \cos \theta$, and $\tan \theta_t = v_v / v_H$.

$$\tan \theta_t = \frac{v_i \sin \theta + gt}{v_i \cos \theta} = \frac{v_i \sin \theta}{v_i \cos \theta} + \frac{gt}{v_i \cos \theta}$$

$$= \tan \theta + \frac{gt}{v_i \cos \theta}$$

Taking $\tan^{-1}$ of both sides completes the proof.

Selected Answers to Discussion Questions

4.1   The hero's ship was already moving rapidly, so he should
        have shut off the engines and coasted, conserving fuel for
        the landing when he must change velocity. The writers
        didn't know the Law of Inertia.

4.5   The pellets rise, all the while slowing down due to drag
        and gravity. They come to rest (at a height of around
        110 m) and then fall back, accelerating at a decreasing
        rate until they reach their terminal speed. Since the
        terminal speed is only about 9 m/s, the shot falls fairly
        harmlessly. By comparison, a .22-caliber bullet, because
        of its large mass, will drop much more rapidly and thus
        can be quite dangerous. Ordinary military bullets have
        still greater terminal speeds and are quite lethal when
        they return to ground level.

4.7   The idea was to fire the shell at a high angle (52°) so
        that it ascended into the stratosphere where the
        atmosphere was rarefied and the drag relatively weak.
        Actually 2 out of the 3.5 minutes of the flight were spent
        sailing through the thin air of the stratosphere.

4.11   When you catch the object, ordinarily your hand is free to
        move and bring it to rest in a relatively long time, which
        means that the force will be small for the same impulse
        and momentum change. Not so if your hand is against a
        table.

4.13   The more massive ball requires the greater impulse to
        bring it up to speed since doing so means a greater
        momentum change. By trying to push or throw an object, an
        astronaut can judge how massive it is and hence how much
        it would weigh. Hitting your finger with a hammer is a
        matter of impulse and momentum-change, and is independent
        of gravity. So yes, it would hurt.

Answers to Multiple Choice Questions

1. a   3. a   5. c   7. c   9. e   11. c   13. b

15. c   17. a   19. c

Solutions to Problems

4.3   The time of flight is t where the vertical distance traveled is $\frac{1}{2}gt^2$.

$1.000\,0$ m $= \frac{1}{2}(9.800\,0$ m/s$^2)t^2$; $t = 0.451\,75$ s.

Then the horizontal distance traveled, which equals the horizontal release distance, is

$$v_H t = (2.213\,6 \text{ m/s})(0.451\,75 \text{ s}) = 1.000\,0 \text{ m}.$$

4.7   From Eq. (3.16) $s_R = -\left(2v_i^2/g\right)\cos\theta\sin\theta =$
$\left[-(2)(405 \text{ m/s})^2/(-9.81 \text{ m/s}^2)\right]\cos 45.0°\sin 45.0° = 16.7$ km.

Obviously in real life there are huge air friction losses. A firing angle less than 45° means a flatter, straighter trajectory, with less distance traveled through air and somewhat less friction loss.

4.19   If the tree is a horizontal distance s_H from the gun, the dart will arrive at $t = s_H/(v_i\cos\theta)$.

At that moment it will have dropped from the line-of-sight by a vertical distance of $\frac{1}{2}gt^2$, but that's the same drop the monkey experiences.

4.23 One force due east, the other two north and south of east at equal angles of $\pm\theta$ such that their N-S components cancel:

$$2 \text{ kN} + (2 \text{ kN}) \cos\theta + (2 \text{ kN}) \cos\theta = 4 \text{ kN}.$$

Then

$$2 \cos\theta = 1, \quad \cos\theta = 1/2, \quad \theta = \pm60°.$$

4.31 The impulses are equal and opposite (compare the two shaded areas), so the final momentum is unchanged: 10 kg·m/s.

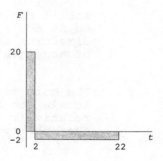

4.39 The initial momentum of the cosmonaut-ship system is zero, so $\mathbf{p}_i = \mathbf{p}_f = 0$.

Then $m_c v_{cf} + m_s v_{sf} = 0$ and $v_s = -(60 \text{ kg})(10 \text{ m/s})/(5000 \text{ kg}) = -0.12 \text{ m/s}$, where the direction of motion of the cosmonaut is positive.

4.49 $F_{av} = mgs_h/s_c$. Since the landing is with both feet, the tolerable force is 100 kN.

Then $100\,000 \text{ N} = (60 \text{ kg})(9.81 \text{ m/s}^2)s_h/(0.010 \text{ m})$, and $s_h = 1.7 \text{ m}$.

4.59 $\mathbf{p}_i = \mathbf{p}_f = 0$, so $(55.0 \text{ kg})v_f = (0.200 \text{ kg})(5.556 \text{ m/s})$.

The skater moves backward at $v_f = 0.020\,2 \text{ m/s}$, in the opposite direction to the snowball.

Answers to Discussion Questions

5.1   The bullet will have momentum and can punch a hole in a silver dollar that has mass and therefore inertia. The silk has so little mass and inertia that the bullet will push it out of the way rather than go through it.

5.5   Direction of net force need not equal direction of motion. A ball can travel horizontally even though the net force on it due to gravity is downward. But since $F = ma$, the direction of net force must be the same as the direction of acceleration.

5.9   The more massive the leg, the more force required to accelerate and decelerate it as the animal runs. The four relatively thin legs on dogs and deer are more efficient than the two thicker legs of humans.

5.11   The water cuts down considerably on the coefficient of friction (see Problem 5.61). By comparison, it would be much harder to skate on dry glass. Skis are waxed to reduce the coefficient of friction. Ice cubes under the marble block would melt allowing the block to slide more easily. The marbles would allow the coffin to be rolled into place rather than slid, and the coefficient of rolling friction is smaller than the coefficient of kinetic friction.

5.13   The locked wheels slide and so have less friction resistance to sideways slippage. Any slight asymmetry and a car with rear wheels locked will spin around so the locked wheels, which are less inhibited, lead the way down the incline. If the brakes cannot be made to engage all at once, it would be best to have the front wheels engage first, to reduce the possibility of the car sluing around 180°. Pumping the brakes allows the wheels, if locked, to reengage the road surface between pumps, which means that the stopping force is static friction rather than the weaker kinetic friction.

Answers to Multiple Choice Questions

1. c   3. d   5. e   7. c   9. b   11. c   13. d

15. d   17. a   19. c

Solutions to Problems

5.3   The smallest force needed is given by

$$F_W = mg = (0.50 \text{ kg})(9.8 \text{ m/s}^2) = 4.9 \text{ N}.$$

This applies only if the frog is being lifted very slowly and so is not subject to significant acceleration.

5.15   The sums of horizontal forces is $\sum F_H = F_T \sin \theta = ma$, where F_T is the tension in the string.

The sum of the vertical forces is $\sum F_V = F_T \cos \theta - F_W = 0$, so $F_T \cos \theta = mg$. Dividing these two equations, $\tan \theta = a/g$.

Since $a = (26.8 \text{ m/s})/(6.8 \text{ s}) = 3.94 \text{ m/s}^2$, $\tan \theta = 0.402$ and $\theta = 22°$.

5.19   The sum of forces parallel to the driveway is $\sum F_\parallel = F_W \sin \theta = ma$, so $mg \sin \theta = ma$, and with $\theta = 20°$, $a = 0.342g$.

From Eq. (3.10),

$$v_f^2 = 2as = 2(0.342)(9.8 \text{ m/s}^2)(20 \text{ m}),$$
$$\text{so } v_f = 12 \text{ m/s}.$$

5.25 $m_1 = m_2 = m$. For the weight carrying the gronch, $F_{W1} - F_T = (m + m_g) g_M - F_T = (m + m_g) a$.

For the other weight, $F_T - F_{W2} = F_T - mg_M = ma$.

Adding the last two equations, $(m + m_g) g_M - mg_M = (2m + m_g) a$ and so $g_M = (2m + m_g) a/m_g$.

5.25   (Cont'd)

$v_f = (1.2 \text{ m})/(3.0 \text{ s}) = 0.40 \text{ m/s}$, so by Eq. (3.10)
$v_f^2 = 2as$ and $a = v_f^2/2s = (0.40 \text{ m/s})^2/2(0.50 \text{ m}) = 0.16 \text{ m/s}^2$.

Then

$g_M = (0.50 \text{ kg} + 0.025 \text{ kg})(0.16 \text{ m/s}^2)/(0.025 \text{ kg}) = 3.4 \text{ m/s}^2$.

5.33   For circular motion, $v^2/R = a_c = 9.0g$, so
$(290 \text{ m/s})^2/9.0(9.8 \text{ m/s}^2) = R$, and
$R = 9.5 \times 10^2 \text{ m}$.

5.41   The sum of forces is $F_N + mg = mv^2/r$.
Here $F_N = F_W/2 = mg/2$, so $v = \sqrt{3gr/2} = 19.2 \text{ m/s}$.

5.51   From Eq. (5.14), $\tan\theta_{max} = \mu_s$, and $\tan 17° = 0.31$.

5.59   Taking the direction of motion as positive,
$\sum F_H = -F_f = -\mu_s F_W = ma$.

Then $a = -\mu_s g$. $v_f = v_i + at$, so $v_i = -at$ and
$t = +v_i/\mu_s g = (27 \text{ m/s})/[0.9(9.8 \text{ m/s}^2)] = 3 \text{ s}$.

The coefficient of kinetic friction does not matter here.

5.67   Taking the direction of motion as positive,
$\sum F_\parallel = F_f - F_W \sin\theta = 0$. $F_f = \mu_s F_N$, so $\mu_s F_N = F_W \sin\theta$.

The normal force on the rear wheels is $(0.43)F_W \cos\theta$,
so $\mu_s(0.43)F_W \cos\theta = F_W \sin\theta$, $0.43\mu_s = \tan\theta$,
$\tan\theta = 0.387$, and $\theta = 21°$. To one significant figure
$\theta = 0.2 \times 10^2$ degrees.

Answers to Discussion Questions

6.1   If the three forces are equal and add up to zero, they must be 120° apart, coplanar and concurrent. One way to see this is to realize that the three vectors, which supposedly add up to zero, must form a closed triangle when placed tip-to-tail. Since the vectors are equal length, the triangle must be equilateral, and this implies the 120° separation. The forces named cannot be the sides of any such triangle since (30 N + 40 N) can at most add vectorially to produce a vector of magnitude 70 N, not 80 N.

6.5   The forearm pivots on the end of the humerus. To raise a load, the biceps contracts, while the triceps stays relaxed; to push down, the triceps contracts, while the biceps relaxes.

6.9   (a) Ordinarily, when touching your toes, you shift the lower portion of your body back as the upper portion leans forward, keeping the *c.g.* over your feet. The wall keeps you from doing this and if you try to touch your toes, you fall forward.

           (b) To stand on your toes, you must move your *c.g.* over your toes so it will be supported. Here, the edge of the door stops you from shifting your mass. If you try to stand on your toes, you fall backward.

6.15   Held tightly together, the boards bend concavely around the fist. The front face of the front board is in compression, while the back board is in tension, being stretched into a curve. The wood along the grain is weak in tension, and the rear board develops a crack at the back face and shatters in two.

6.17   Your arms have a downward force acting on them at the hands, from the weights, and an equal and opposite upward force acting on them at the shoulders. So your arms are in tension. Your legs, on the other hand, have a downward force acting on them at the hip joints from the weight of your upper body and the weights in your hands. Your legs

6.17   (Con't)

have an equal and opposite upward force acting on them at the soles of your feet from the earth. So your legs are in compression.

Answers to Multiple Choice Questions

1. d   3. d   5. c   7. e   9. b   11. a   13. e

15. d   17. d   19. c

Solutions to Problems

6.1   (a) The force on the hook is a downward force of 200 + 50 N = 250 N.

        (b) The force on the bottom of the chain is a downward force of 200 N.

6.11   The arrangement is static, so the sum of forces acting on the ring in the middle must be zero. Then

$+\uparrow \sum F_v = (15.0 \text{ N}) \cos 45.0° + (31.0 \text{ N}) \cos 20.0° - F_{W3} = 0,$

thus F_{W3} = 39.7 N.

In equilibrium, not only $+\uparrow \sum F_v$ but also $\overset{+}{\underset{\rightarrow}{}} \sum F_H$ must equal zero, and indeed $-(15.0 \text{ N}) \sin 45.0° + (31.0 \text{ N}) \sin 20.0° = 0.$

6.17   Assuming the arrangement is in equilibrium, the sum of forces acting on mass m must be zero. Then

$+\uparrow \sum F_v = 2 F_T \sin 20° - F_W = 0,$

where F_T = (1.0 kg)g and F_W is the weight of mass m. Hence F_W = 6.7 N.

6.23 F_{T1} = (80.0 kg)g = 784.5 N and
F_{T2} = (10.0 kg)g = 98.07 N.

Since $\overset{+}{\rightarrow}\sum F_H$ = 784.5 N - F_f - 98.07 N = 0,
F_f = 686 N, there is a leftward static friction force of
686 N acting on the box.

The fact that the fly's weight is enough to cause slippage
means that $|F_f| \approx \mu_s F_N$. F_N = (75.0 kg)g, so

$$\mu_s = \frac{686 \text{ N}}{(75.0 \text{ kg}) g} = 0.93.$$

It makes no sense to give μ_s to more than 2 significant
figures.

6.27   The moment-arm is 1.00 m, so the torque is given by

$$\tau_0 = (1.00 \text{ m})(100 \text{ N}) = 100 \text{ N·m}.$$

6.37   Take the support point of the stand to be point A.
The net torque acting on the beam must be zero, so

$\overset{\curvearrowright}{+}\sum \tau_A$ = - (2.0 kg)g(0.10 m) - F_{RB}(0.20 m)
+ (1.0 kg)g(0.40 m) = 0,

where F_{RB} is the magnitude of the reaction force of the
scale acting on the beam. The scale reads

$$\frac{F_{RB}}{g} = \frac{(1.0 \text{ kg}) g}{g} = 1.0 \text{ kg}.$$

6.51   The sum of horizontal forces is zero, so
$\overset{+}{\rightarrow}\sum F_H$ = $F_{T2} \cos 60.0°$ - (100 N) $\cos 60.0°$ = 0,
which means F_{T2} = 100 N. Scale 2 reads 100 N.

The sum of the vertical forces is zero, so
$+\uparrow\sum F_V$ = 2(100 N) $\sin 60.0°$ - F_W = 0,
and F_W = 173 N.

The vertical line passing through the *c.g.* is the
line-of-action of $\mathbf{F}_W$, so it passes through the point of
intersection of the lines-of-action of the two tension
forces.

6.61   The net torque about the toes is zero, so

$(+\sum \tau = 2\,F_{Nh}\,(1.50\text{ m}) - (65\text{ kg})\,g\,(1.00\text{ m}) = 0$,

where F_{Nh} is the force on each hand. Then $F_{Nh} = 212$ N = 0.21 kN.

The net force acting on the woman's body is also zero, so $+\uparrow\sum F_V = 2\,(212\text{ N}) + 2\,F_{Nt} - (65\text{ kg})\,g = 0$ where F_{Nt} is the force on each foot at the toes. So $F_{Nt} = 107$ N = 0.11 kN.

Answers to Discussion Questions

7.3   Yes, this is possible. The gravitational field inside a uniform spherical shell of mass is zero. [See Fig.(7.5).] But the mass must be completely symmetrical. There is no shielding of the inside, and new arrivals would be detected gravitationally.

7.5   When the craft is tilted, the little mass at the end of the boom and the main portion of the satellite both tend to rotate about the craft's *c.g.*, lying between them. The Earth's gravity drops off as $1/r^2$, where r is the distance from the earth's center. Given a small deflection, the mass at the bottom of the boom experiences a greater field and develops a greater torque, pulling the boom back to vertical. The craft will thereby remain stable pointing downward all the time.

7.7   The Earth rotates counterclockwise as seen looking down onto the North Pole, which means that any rocket on the launch pad is already moving eastward at the speed of the Earth even before lift-off. The closer to the equator, the greater the speed of a point on the Earth's surface. So, launch from a southerly location and launch eastward and you get as much benefit from the planet's rotation as possible. Florida and southern Texas are the southernmost areas of the contiguous 48 states. Placing the launch site on Florida's eastern coast minimizes the chance that a failed launch falls in a populated area.

7.11   The rocket fires with a forward thrust and that carries it into an elongated elliptical orbit. The additional speed allows it to move outward beyond the circular orbit. It slows down as it moves away from the center of force. As it reaches its most distant point (just before it would begin to fall back inward), the rocket fires again, giving it enough speed to satisfy Eq.(7.7) and go into a large circular orbit. Despite all maneuvering, the final orbital speed is less than the initial orbital speed.

7.15   These antennas pick up TV signals from satellites parked
        in geosynchronous orbits. In the Northern Hemisphere,
        these are all seen in the southern part of the sky, since
        they are over the equator. So the antennas face generally
        southward. Each satellite is fixed at a known location in
        space, so you need only decide which one you would like to
        tune in.

7.17   Since gravity increases as the mass of a body increases,
        there will come a point where the gravitational force
        tending to squash a mass exceeds all structural forces
        keeping it from deforming. Beyond that limit, which is
        around 1000 km in diameter, a typical solid celestial body
        will become essentially spherical. Apparently, Hyperion is
        just too small.

7.19   Any point on the plane of the orbit is at distances from
        the Sun and Earth such that the magnitude of the
        gravitational attraction to each celestial body is the
        same. The resultant of the two opposing forces will
        therefore only have a component in that perpendicular
        plane, and it will always point toward the Earth-Sun axis.
        Thus, it will act as a central force, and the satellite
        will be held in orbit as if there were a central
        gravitating body.

Answers to Multiple Choice Questions

   1. c   3. d   5. c   7. c   9. b   11. d   13. c

15. b

Solutions to Problems

7.1 $F_W = GmM_e/R^2$. Doubling the mass and the distance yields

$$F_W' = G(2m)M_e/(2R_e)^2 = F_W/2.$$

7.17   For $r \geq R_e$, $mg_e = GmM_e/r^2$.
        $g_0 = GM_e/R_e^2$, hence

$$g_\oplus = GM_\oplus/r^2$$
$$= (GM_\oplus/R_\oplus^2)(R_\oplus^2/r^2)$$
$$= g_0(R_\oplus^2/r^2)$$
$$= g_0(R_\oplus/r)^2.$$

7.25   Setting the field strength acting on a small particle on the star's surface equal to the acceleration needed to keep the particle on the surface as the star spins, one gets $GM_n/R_n^2 = v^2/R_n$.

For a particle on the equator of a spinning sphere, then,

$$[G(4/3)\pi R_n^3 \rho]/R_n^2 = (2\pi R_n/T)^2/R_n, \text{ or } (G\rho/3\pi)^{1/2} = 1/T.$$

Then $T = 1 \times 10^{-3}$ s.

7.31   The Orbiter's distance from the moon's center is 1738 + 62 km = 1800 km. The orbital speed is then given by Eq.(7.8):

$$v_0 = \sqrt{GM_\mathfrak{c}/r_\mathfrak{c}} \approx [(6.67 \times 10^{-11} \text{ N·m}^2\text{·kg}^{-2}) \times$$
$$(7.35 \times 10^{22} \text{ kg})/(18.0 \times 10^5 \text{ m})]^{1/2} = 1.65 \times 10^3 \text{ m/s}.$$

7.39   Plugging the sun's mass and the earth's distance for the sun's center into Eq.(7.9), one gets

$$g_\odot = GM_\odot/r_\odot^2 = G(2.0 \times 10^{30} \text{ kg})/(1.5 \times 10^{11} \text{ m})^2$$
$$= 5.9 \times 10^{-3} \text{ m/s}^2.$$

7.47   Since the gravitational influence of a uniform shell on a point in its interior is zero, the field intensity at the point in question is the same as it would be on the surface of sphere having half the radius and hence one-eighth the mass. Thus,

$$g = \frac{G(M/8)}{(R/2)^2} = \frac{GM}{2R^2}.$$

Answers to Discussion Questions

8.1   Because the outside wheels have farther to go, they turn
        faster. A rigid common axle would be subjected to twisting
        forces at every turn and would soon break. This is why a
        differential gear is mounted between pairs of wheels.

8.5   A small force acting with a large enough moment-arm may
        produce more torque, and hence greater angular
        acceleration, than a large force acting with a small
        moment-arm. No, nonzero angular momentum does not entail
        nonzero net torque. In a frictionless environment, a body
        may continue spinning indefinitely without any net torque
        acting on it.

8.7   It must be zero. If it were not, there would be a nonzero
        net torque about the sphere's center. But L is constant,
        so $\Delta L = 0$ and $\tau = 0$.

8.9   The fact that the line-of-action of the force does not
        pass through the $c.m.$ results in a torque; L, ω, and τ are
        up out of the page. There will be a slight rise of the
        $c.m.$ and then a ballistic flight. She gains angular
        momentum and will flip over.

8.13 L in the ball-person system is not conserved because there
        is a friction force via the ground on the thrower's feet.
        Enlarging the system to include the Earth conserves
        angular momentum, and so the Earth's motion must change
        accordingly. The person would be pushed backward and would
        also rotate, with the upper body moving away from the
        ball.

8.15   At a constant speed, the
        forward propeller reaction
        force **F**, which is beneath
        the $c.m.$, equals the
        friction force arising
        from the air and water,
        just as the weight equals
        the upward buoyant force,
        and so $\sum \mathbf{F} = 0$. At any

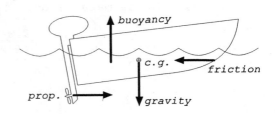

8.15   (Cont'd)

moment (neglecting friction and any vertical component from the propeller force), there is a balance between the torque from the propeller force and the buoyant force. The greater **F** is made, the more the boat noses up and the farther toward the rear (away from the *c.m.*) the point of action of the buoyant force moves, thereby increasing its torque until equilibrium is reached again.

8.19   His angular speed will remain unchanged, since he has been subjected to no net torque, and the moment of inertia *of his body* has not changed. Nor has the angular momentum of the student-weight system changed. The weights continue to contribute angular momentum as they roll across the floor.

Answers to Multiple Choice Questions

1. b   3. e   5. b   7. c   9. b   11. b   13. d

15. a   17. d   19. b

Solutions to Problems

8.7   The distance between adjacent lines is given by
l = (0.30 m)/525 = 0.571 × 10^{-3} m.

One minute of arc equals $\frac{1}{60}°$ = 0.016 67° = 0.000 29 rad.

Since $l = r\theta$, the distance r at which adjacent lines will become indistinguishable is
(0.571 × 10^{-3} m)/(0.000 29 rad) = 1.96 m = 2.0 m.

8.15 $\omega_f = \alpha t$, so (500 s^{-1})(2π rad) = α(30 s).
α = 105 rad/s^2.

8.23   Converting ft/s to mi/h

$\omega = v/R = (1.467$ ft/s per mi/h$)(20$ mi/h$)/(1000$ ft$)$
$= 0.029$ rad/s.

$a_c = R\omega^2 = (1000$ ft$)(0.029$ rad/s$)^2 = 0.86$ ft/s$^2 = 0.26$ m/s^2.

8.27   The speed of a point on the belt is $R_A\omega_A = R_C\omega_C$.
The speed of a point on cable E, and so of the suspended
body, is given by $v_E = R_D\omega_D = R_D\omega_C$.

So $v_E = R_D(R_A\omega_A/R_C) = (0.4$ m$)(0.3$ m $\times 2\pi$ rad $\times 1$ min^{-1} $\times$
$1/60$ min/s$)/(0.1$ m$) = 0.13$ m/s, and to one significant
figure, $v_E = 0.1$ m/s, upward.

8.39   33⅓ rpm = 33⅓ $\times \dfrac{2\pi}{60}$ rad/s = 3.49 rad/s.

The speed of a point on the belt is $v_B = R_p\omega_p = R_t\omega_t$ where p
stands for pulley and t for turntable.

When $\omega_t = 3.49$ rad/s, then,
$\omega_p = (15$ cm$)(3.49$ rad/s$)/(1.0$ cm$) = 52.4$ rad/s.

To reach this angular speed in 6.0 s, the pulley must
undergo an angular acceleration of $(52.4/6.0)$ rad/s^2 =
8.73 rad/s^2 = 8.7 rad/s^2. The angular distance traveled by
the pulley in that time is given by

$\theta = \frac{1}{2}\alpha t^2 = \frac{1}{2}(8.73$ rad/s$^2)(36$ s$^2) = 157$ rad = 25 rev

8.41   Letting ω_i be the angular speed before an increase in
period of ΔT and ω_f the angular speed after the increase,
$\omega_i = 2\pi/T$, $\omega_f = 2\pi/(T + \Delta T)$, where T represents one 24-hour
day.

Then $\Delta\omega = -2\pi/T + 2\pi/(T + \Delta T) = -2\pi\Delta T/T(T + \Delta T) \approx -2\pi\Delta T/T^2$.

$\alpha = \Delta\omega/\Delta t$, which, when Δt is one day, yields
$\alpha = \Delta\omega/T = -2\pi\Delta T/T^3$
$= (-2\pi$ rad $)(25 \times 10^{-9}$ s$)/(24$ h $\times 60$ min/h $\times 60$ s/min$)^3$
$= -2.4 \times 10^{-22}$ rad/s^2.

For constant acceleration,
$\Delta\omega = \alpha t$
$= (-2.4 \times 10^{-22}$ rad/s$^2)(1 \times 10^9$ y $\times 364.25$ d/y $\times 86,400$ s/d$)$
$= -7.6 \times 10^{-6}$ rad/s.

8.41   (Cont'd)

$\omega_f = \omega_i + \alpha t$.
Adding $\Delta\omega$ to the current ω of $2\pi/(24 \times 60 \times 60)$ rad/s
yields a new ω of 6.51×10^{-5} rad/s, corresponding to a
period of $\dfrac{2\pi}{\omega} = 9.65 \times 10^4$ s $= 26.8$ h $= 27$ h.

8.55   The sum of forces on the mass is given by
$+\downarrow \sum F_V = ma = F_W - F_T$, where F_W is the weight of the mass and
F_T is the tension in cord.

The sum of torques on the pulley is given by
$\overset{\curvearrowleft}{+} \sum \tau = I\alpha = F_T R$. Since $a = R\alpha$, $F_T = I\alpha/R = Ia/R^2$.

Then $F_W - Ia/R^2 = ma$, $a(m + I/R^2) = mg$,
and $a = mg/(m + I/R^2)$.

8.63   At a minimum, $\mu_s = F_f/F_N$.

From Eq. (8.35), $F_f = (2/7)mg\sin\theta$, so the minimum value of
μ_s is $(2/7)mg\sin\theta/mg\cos\theta = (2/7)\tan\theta$.

8.69   The distances of each of the 100 parts from O are given by
$r_1 = l/200$, $r_2 = l/100 + l/200$, $r_3 = 2l/100 + l/200$, ... ,
$r_n = 99l/100 + l/200$.

Then

$$I = \sum mr^2$$
$$= (M/100)[(l/200)^2 + (3l/200)^2$$
$$+ (5l/200)^2 + ... + (199l/200)^2]$$
$$= [Ml^2/(4 \times 10^6)][1^2 + 3^2 + 5^2 + ... 199^2].$$

Since

$$[1^2 + 3^2 + 5^2 + ... 199^2] = (100/3)(201)(199),$$
$$I = Ml^2(100/3)(201)(199)/(4 \times 10^6).$$

Hence, $I \approx (1/3)Ml^2$.

Answers to Discussion Questions

9.1   No work is done if the rocket doesn't move. The rocket
        pushes down on the exhaust gas and the gas pushes up on
        the rocket, accelerating it. The work done on the rocket
        increases with time as it moves faster and therefore
        farther per second. With respect to the ground, the gas
        initially gets most of the KE. As the vehicle speeds up,
        more energy is transferred to it. All the KE, for both
        rocket and exhaust, comes from the chemical energy of the
        fuel. What makes this different from swimming is that the
        exhaust gas does positive work on the rocket, since the
        underside of the rocket is moving in the direction of
        travel. Water does no positive work on a swimmer, since
        the propelling surfaces of the hands and legs move *against*
        the direction of travel.

9.3   (a) He is overcoming air drag.

        (b) The reaction of the ground on him is the propelling
             force. If his feet slide back, the forward-directed
             force of the-floor-on-him will do negative work on
             him. He will do work heating the floor.

        (c) If his feet do not slip, he does no work on the floor
             and the floor does no work on him.

        (d) His KE comes from his food.

9.5   (a) Yes, he did work on the bike and provided the energy
             needed for the trip.

        (b) The external force is the reaction force of the
             ground.

        (c) No, with no slipping there's no relative motion
             between the tire and ground, and no work is done on
             the bike by the road. Even when there is distortion of
             the tire and road, the work is done on the road, since
             the rider is the only energy source.

9.7   The speeds at the beginning and end of each track are the
        same, but the ball moving along the depression wins the
        race. The component of the weight of the ball acts in the
        direction of motion as the ball descends and opposite to
        it as it ascends.

9.7   (Cont'd)

That means that the ball is first accelerated above the
initial horizontal speed and then decelerated back to its
initial horizontal speed. The forward speed of the ball in
the depression, then, will always be equal to or greater
than the other ball's.

9.9   Work is energy transferred to or from a system via the
application of a force acting over a distance.
Historically, it got its name before the concept of energy
was formalized, and so it continued to be treated as if it
were something other than energy. But defining energy as
the ability to do work means defining energy as the
ability to transfer energy. It's just as circular as
saying energy is the ability to produce *vis viva* (KE). It
makes sense, since work and PE both have the ability to
produce *vis viva*, but it's clearly circular.

9.11   At a constant speed there is a force, but it's
perpendicular to *s*. Hence there is no work done and no
change in PE or KE. When the ball has a tangential
acceleration there is a tangential force — work done — and
an increase (or decrease) in KE. A tangential component
can exist when the string leads the ball, making an angle
with *v* of less than 90°. PE remains the same.

9.13   (a) Gravitational-PE is greatest at the top of the first
             hill.

        (b) For highest possible maximum speed, you want the
             lowest, steepest drop right at the beginning of the
             ride.

        (c) Gravitational-PE cannot return to maximum, since
             energy has been expended through friction.

        (d) Maximum KE can be attained more than once, by making
             each valley lower than the one before by just enough
             to regain the KE lost to friction.

        (e) No.

        (f) All hills have to be lower than the first.

9.17   The friction force accelerates the crate such that
$W = \Delta KE = \frac{1}{2}mv^2$. The work is positive. The work overcomes
inertia. The energy appears as KE. During stopping, the
friction force is in the opposite direction to the

displacement and does negative work, converting KE to
thermal energy released as heat by the truck brakes.

Answers to Multiple Choice Questions

1. a   3. a   5. a   7. c   9. c   11. a   13. a

15. c   17. e   19. c   21. d

Solutions to Problems

9.13   There are 5 ropes supporting the load, hence 5.0 m of rope
will have to be pulled out.

9.17   The work is given by

$$W = \mu_r mg (\cos \theta) s = (4.9 \text{ N}) (\cos 10°) (25 \text{ m}) = 1.2 \times 10^2 \text{ J}.$$

9.25   The person's power consumption is

$(0.40 \text{ liter/min}) (2.0 \times 10^4 \text{ J/liter}) = 8.0 \times 10^3 \text{ J/min}$
$= 1.3 \times 10^2 \text{ W}.$

$BMR = (1.3 \times 10^3 \text{ W}) / (1.8 \text{ m}^2) = 74 \text{ W/m}^2.$

9.31   From Conservation of Momentum, $v_G = m_b v_b / m_G = 0.98 \text{ m/s}.$

Then $KE_G = \frac{1}{2} m_G v_G^2 = 0.95 \text{ J}$ while
$KE_b = \frac{1}{2} m_b v_b^2 = 0.29 \text{ kJ}.$

9.41   The 100-N weight doesn't move.

(a) $W = (10 \text{ N}) (10 \text{ m}) = 100 \text{ J}.$

(b) Since the weight is too heavy to rise, no rope ends up
on the floor.

(c) $\Delta PE = +100 \text{ J}.$

9.47   From Table 8.3, the rotational kinetic energy for a
        uniform sphere is
        $(\frac{1}{2})(2/5)(mR^2)(v/R)^2 = (2/5)(\frac{1}{2})(mv^2)$,
        so the total kinetic energy is $(7/5)(\frac{1}{2})(mv^2)$.

        This is enough to raise the sphere a distance of
        $(7/5)(\frac{1}{2})(mv^2)/(mg)$ m = 1.79 m vertically, which corresponds
        to a distance up the slope of $1.79/\sin 20°$ m = 5.2 m.

9.51 $\Delta PE = mg\Delta h = (55.0 \text{ kg})(9.80 \text{ m/s}^2)(9.00 \text{ m}) = 4.85 \times 10^3$ J.

        This is the energy stored in the trampoline. With no
        losses, she will rise 9.00 m or back up to 10.0 m above
        ground.

9.59   Compare the two values of kinetic energy at escape speeds:

        $$\left[\frac{1}{2}mv_{esc}^2 (\text{Moon})\right]/\left[\frac{1}{2}mv_{esc}^2(\text{Earth})\right] = v_{esc}^2(\text{Moon})/v_{esc}^2(\text{Earth})$$
        $$= (2.4 \text{ km/s})^2/(11.2 \text{ km/s})^2 = 0.046 \approx 5\%.$$

9.61   Since $v_{esc} = \sqrt{2}\, v_o$, the escape speed is $1500\sqrt{2} = 2121$ m/s.

9.71   The velocity vector v_i of the moving ball has two
        components, $v_{i\parallel}$ along the line-of-centers and $v_{i\perp}$
        perpendicular to that. The result is as if the two motions
        occurred independently—the perpendicular motion
        constitutes a miss while the parallel motion transfers all
        that momentum to the target ball, which sails off along
        the line-of-centers while the incident ball, no longer
        possessing momentum in that direction, moves away
        perpendicular to the line-of-centers.

Answers to Discussion Questions

10.1   Aluminum — foil, pots; tungsten — light bulbs, in steel knife blades; carbon — in wood, pencil lead, diamond; mercury — in fluorescent bulbs, silent switches, thermometers; zinc — paint pigment, ointments, in brass; chlorine — bleach, drinking water, salt; copper — pots, wire, jewelry, in brass and bronze; sodium — table salt, Alka-Seltzer; iron — blood, raisins, nails, pots; lead — plumbing, solder, old paint; magnesium — griddles, antacid pills, sparklers; cobalt — blue dyes and pigments; oxygen — air, water, rust; chromium — in steel, plated on toasters and car bumpers; nickel — alloyed in coins, in common Alnico magnets.

10.3   (a)   Low-carbon steel is the most ductile.

         (b)   Tempered steel is the most brittle.

         (c)   At low strain, all curves have the same slope, hence the materials have the same Y.

         (d)   Tempered steel.

         (e)   Tempered steel.

         (f)   Low-carbon steel.

         (g)   Since Y is the same for all, elastic stiffness is the same for all.

         (h)   Low-carbon steel.

10.5   (a)   The area under the force-displacement curve is the work done on the sample, the energy mechanically entered into the material in the process of distorting it.

         (b)   Since stress is force over area, and strain is displacement over length, the product of the two is the strain energy divided by the volume. It is, in other words, the amount of work per unit volume required to distort the sample. It has the units of J/m^3.

         (c)   The colored area corresponds to the increased internal energy per unit volume of the sample.

10.7   Both ropes break when the stress exceeds a certain value
        that is the same for each, so they can carry the same
        load. But more work will have to be done to break the
        longer rope (which elongates more) than the shorter one.

10.9   The stress in the broad base, which carries the entire
        load, is less, which is one reason why ancient walls
        tapered upward. The stress on the base of a column is
        $Ah\rho g/A = h\rho g$, so the maximum height is related to the
        ultimate compressive strength divided by ρg.

        A column of marble ($\rho = 2.7 \times 10^3$ kg/m^3) which is 4.2 km
        high would be too heavy for its base, which can only
        support a column up to 4.15 km high. A granite column
        could be up to 9.06 km high.

10.11   Elephants have to walk around very carefully (they don't
        do much jumping). The compressive strength (Table 1) of a
        horse femur is only 145 MPa. We have bred horses to the
        point where they are almost too big for their bones — at
        least with all the jumping and running we demand of them.

10.13   More work goes into the rubber than comes out, and the
        difference (the area of the colored region in the figure)
        appears mostly as thermal energy. Rubber is therefore
        useful to convert unwanted mechanical energy into thermal
        energy and is often used to damp vibrations and shocks.
        When a rubber band is worked cyclically, it will get
        noticeably warmer.

10.15   Human tendon has to be able to store a great deal of
        energy without permanently distorting — it's highly
        resilient, twice as much as spring steel. Bridges and
        springs must also be resilient; they must recover after
        being compressed or stretched. Glass has very little
        ductility. The area under its stress-strain curve is small
        and it has little toughness — that's why a glass or china
        plate will shatter when dropped. A bow or pole should be
        resilient.

Answers to Multiple Choice Questions

1. d   3. d   5. b   7. d   9. c   11. b   13. b

15. a   17. c   19. b

Solutions to Problems

10.3   Since 1 lb is equivalent to 0.453 6 kg,

$$2000 \text{ lb/in.}^3 = (907.2 \text{ kg})/(2.54 \times 10^{-2} \text{ m/in.})^3$$
$$= (907.2 \text{ kg})/(16.39 \times 10^{-6} \text{ m}^3)$$
$$= 6 \times 10^7 \text{ kg/m}^3.$$

10.15 $N_A \rho/M_m = (6.022 \times 10^{26}$ molecules/kilomole)
$$\times (1.000 \times 10^3 \text{ kg/m}^3)/(18 \text{ kg/kilomole})$$
$$= 3.3 \times 10^{28} \text{ molecules/m}^3.$$

10.19   From the chemical formula its molecular mass is

$$2(12 \text{ u}) + 5(1 \text{ u}) + 16 \text{ u} + 1 \text{ u} = 46 \text{ u}.$$

Hence from Eq. (10.2),
$$L = [(46 \text{ g/mol})/(6.022 \times 10^{23} \text{ molecules/mol}) \times$$
$$(0.789 \text{ g/cm}^3)]^{1/3}$$
$$= 4.59 \times 10^{-8} \text{ cm}$$
$$= 0.46 \text{ nm}.$$

10.27   The strain is given by

$$\epsilon = \Delta L/L_0 = (1.5 \times 10^{-2} \text{ m})/(10 \text{ m}) = 0.15\%.$$

10.35   The pressure is given by

$$P = F/A = (10 \text{ N})/(0.20 \text{ m} \times 0.30 \text{ m}) = 1.7 \times 10^2 \text{ N/m}^2.$$

10.39   The potential energy stored is $\Delta PE_e = \frac{1}{2}ks^2$, so

$$k = 2(3.2 \times 10^{-19} \text{ J})/(0.2 \times 10^{-9} \text{ m})^2 = 16 \text{ J/m}^2.$$

10.43   The new cross-sectional area is four times the old, hence
the breaking load is 4(736 3 N) = 29.5 kN.

10.47   25 000 lb = 111 kN, and if this represents the cable's yield point, it equals $A\sigma = \pi R^2 \sigma$.

Then $(111 \text{ kN})/\pi (345 \text{ MPa}) = R^2$. R must exceed 1.01 cm, and so the diameter must exceed 2.02 cm.

The maximum strain is given by

$$\epsilon = \sigma/Y = (345 \text{ MPa})/(200 \text{ GPa}) = 0.173\%.$$

10.51   The total area being sheared is

$$5A = 5\pi R^2 = 6.28 \times 10^{-3} \text{ m}^2;$$

$\sigma = F/A$, so $F = \sigma A = (145 \text{ MPa})(6.28 \times 10^{-3} \text{ m}^2) = 0.911 \text{ MN}.$

Answers to Discussion Questions

11.1   (a) The reading will increase as the effective g
             increases.

        (b) Zero.

11.5   Work is done filling the balloon and displacing the
        surrounding atmosphere. *Gravitational*-PE is stored in the
        balloon-atmosphere system. As the balloon ascends, the
        atmosphere descends.

11.7   (a) Nothing — you now displace one-cup's weight more of
             water, so the level stays the same.

        (b) The bust floats partially submerged, displacing its
             weight of water. When it was aboard, it was displacing
             its weight as well, so the level again does not
             change.

        (c) Nothing.

        (d) The level drops, since the raft now displaces slightly
             less water.

11.11   The difference between P_1 and P_T keeps the liquid in the
        column, hence $P_T = P_1 - \rho gh$ and when $P_1 = 0$, $P_T < 0$.

11.13   The soap drops the surface tension in the middle and the
        surface pulls out to the periphery, as if you had punched
        a hole in a stretched rubber sheet.

11.15   Air streaming over it asymmetrically (the ball drops a
        little, allowing most of the air to rush over it) produces
        a pressure drop above the ball (where a lot of air is
        moving rapidly), resulting in lift.

11.17   Air beneath the ball will move more rapidly than above;
        the resulting pressure drop below the ball will cause it
        to sink more swiftly than if it was not spinning. On the
        ship, the rotating columns acted like two wings

11.17   (Cont'd)

with their leading edges turned into the wind. By reversing the columns' direction of rotation as necessary, the ship's crew could get forward power from any wind except a direct head- or tailwind.

11.19   To have lift, the wing must have air circulating around it, which means angular momentum. Somehow, an equal and opposite amount of angular momentum will be imparted to the air-plane system and that's done nicely with the creation of a starting vortex revolving in the opposite direction. If the wing is stopped, the circulating air will generate another vortex equal and opposite to the starting vortex.

Answers to Multiple Choice Questions

1. b   3. d   5. a   7. a   9. a   11. d   13. b

15. c   17. c   19. d

Solutions to Problems

11.5 $1.000 \text{ lb/in.}^2 = (4.448\,23 \text{ N})/(1 \text{ in.} \times 2.540 \times 10^{-2} \text{ m/in.})^2$
$= 6.895 \times 10^3 \text{ N/m}^2.$

11.13 $P_i V_i = P_f V_f$, so

$$(1.013 \times 10^5 \text{ Pa})(10 \text{ cm}^3) = P_f (2.5 \text{ cm}^3).$$

$P_f = 4.1 \times 10^5$ Pa.

11.21   The gauge pressure necessary to raise a column of water 1.1 m high is given by

$$P_G = \rho g h = -(1.00 \times 10^3 \text{ kg/m}^3)(9.8 \text{ m/s}^2)(1.1 \text{ m})$$
$$= -1.1 \times 10^4 \text{ Pa.}$$

11.27   The weight of the sea water displaced (of volume V_W)

equals the weight of the entire berg (of volume V_i), hence

$$V_w (1.025 \times 10^3 \text{ kg/m}^3) g = V_i (0.92 \times 10^3 \text{ kg/m}^3) g.$$

V_w/V_i = 89.8% or 90% is submerged and 10% is visible.
For fresh water, V_w/V_i = 92% and 8% is above.

11.35   A volume of 2.54 cm section is

$$(3.05 \text{ m}) (6.10 \text{ m}) (2.54 \times 10^{-2} \text{ m}) = 0.473 \text{ m}^3,$$

and that much water weighs

$$(0.473 \text{ m}^3) (1.00 \times 10^3 \text{ kg/m}^3) g = 4.63 \text{ kN}$$

which is the load taken on per 2.54 cm of settle.

The raft weighs

$$(5.67 \text{ m}^3) (0.50 \times 10^3 \text{ kg/m}^3) g = 27.8 \text{ kN}$$

which equals a weight in water of $V_w (1.00 \times 10^3 \text{ kg/m}^3) g$.
Then V_w = 2.84 m³, which amounts to a depth of

$$(2.84 \text{ m}^3) / (3.05 \text{ m}) (6.10 \text{ m}) = 0.15 \text{ m}.$$

11.49 $\Delta P = \rho g h$, which for the density of blood equals

$$(1.05 \times 10^3 \text{ kg/m}^3) (9.81 \text{ m/s}^2) (0.85 \text{ m})$$
$$= 8.75 \text{ kPa or } 8.8 \text{ kPa}.$$

The pressure at the heart is 13.3 kPa, so the reduction is
significant.

11.53   From Torticelli's result, the horizontal flow speed is
$\sqrt{2gh}$.

To fall a height $y = \frac{1}{2}gt^2$ takes a time $t = \sqrt{2y/g}$.

During that time, the flow will travel a horizontal
distance and hit the floor at $x = \sqrt{2gh}\sqrt{2y/g} = 2\sqrt{yh}$.

11.57 From Bernoulli's Equation, $P_0 + \frac{1}{2}\rho v^2 = P$.

Then $\Delta P = \frac{1}{2}\rho v^2$, and $v = \sqrt{2 \, \Delta P/\rho}$.

11.59 From Eq. (11.22), $P_p - P_t = \frac{1}{2}\rho v_t^2 (A_p^2 - A_t^2)/A_p^2$.

Since $\Delta P = \rho g \Delta y$ and from the Continuity Equation

$A_p v_p = A_t v_t$, $v_t^2/A_p^2 = v_p^2/A_t^2$ and

$$\rho g \Delta y = \frac{1}{2}\rho v_p^2 (A_p^2 - A_t^2)/A_t^2.$$

Then $v_p = \sqrt{2g\Delta y/[(A_p^2/A_t^2) - 1]}$.

11.61 From Bernoulli's Equation, $P_i - P_A = \frac{1}{2}\rho v^2$.

With units of pressure converted to MPa

$$(3.79 \text{ MPa} - 0.101 \text{ MPa})2/\rho = v^2.$$

With the given ρ, $v = 82$ m/s, or about 180 mph.

11.65 From Bernoulli's Equation, $P_1 + \frac{1}{2}\rho v_1^2 + \rho g y_1 = P_2 + \frac{1}{2}\rho v_2^2 + \rho g y_2$.
Using gauge pressure and taking $y = 0$ at the hole,

$$0 + 0 + \rho g(H + h) = \rho g h + \frac{1}{2}\rho v_2^2 + 0,$$

so $v_2 = \sqrt{2gH}$.

Answers to Discussion Questions

12.1   When θ is a maximum, $v = 0$, as is KE,
        but a is a maximum, as is PE.
        When $\theta = 0$, $a = 0$, as is PE,
        but v is maximum, as is KE.

12.3   The force tending to stretch the spring is a maximum twice
        during each cycle of the pendulum (viz., when it's
        vertical). Thus the frequency of the spring oscillation
        should be made equal to twice the pendulum frequency. See
        W.R. Mellen, <u>Phys. Teach.</u> **32**, 122 (1994).

12.5   Increasing the mass will decrease the natural frequency.

12.7   The parachute will swing like a pendulum, and as the
        vortex frequency matches the pendulum frequency, the
        resulting resonance will send the rider wildly swinging.

12.9   Like the bob on a spring, the greater the mass the smaller
        the frequency.

12.11   At any moment (when at a distance r from the center of
        Mongo), the weight of the ball depends on the mass M of
        material remaining "beneath" it. Only the mass of the
        sphere of matter of radius r attracts the ball. $F_G \propto mM/r^2$,
        where M is proportional to the volume, which goes as r^3.
        Hence, $F_G \propto r$ and the motion must be SHM.

12.13   The waves vanish, which certainly means there is no
        elastic-PE stored in the undistorted rope. All the energy
        is kinetic; though undisplaced at that very instant, the
        various segments of the rope in the region of overlap are
        nonetheless moving vertically.

12.15   The speed of the wavepulse varies with the square root of
        the tension, which, in turn, is determined by the load and
        the weight of the string itself. The tension increases all
        the way up to the point of support because of the linear
        mass density of the string. Hence, the speed increases as
        the wave rises. Air friction and internal losses will

12.15   (Cont'd)

convert some of the energy of the wave to thermal energy
and its amplitude will diminish.

12.17   The materials are getting progressively more rigid as we
go from clay down the table. (Compare the softness of clay
to the hardness of granite.) Accordingly, we can expect
that the internal restoring force will increase as the
interatomic force increases, and so the speed of a
compression wave will increase. Of course, an increase in
rigidity corresponds to an increase in Young's Modulus.

12.19

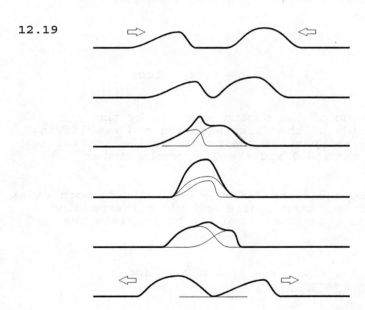

Answers to Multiple Choice Questions

1. b   3. a   5. c   7. c   9. d   11. c   13. a

15. d   17. d   19. b   21. c

Solutions to Problems

12.3   The motion is SHM along a line.

The frequency is given by

$$f = (78 \text{ rot/min})/(60 \text{ s/min}) = 1.3 \text{ Hz}.$$

The angular frequency, ω, is $2\pi f = 2.6\pi$ rad/s.

12.19   Since $F = mg = ky$, $m = 2.00$ kg and $y = 0.025\ 0$ m,
        $k = 784$ N/m.

$$f = (1/2\pi)\sqrt{k/m} = 3.15 \text{ Hz}.$$

12.27   From Eq.(12.5) and (12.7) at $t = 0$, $x_i = A\cos\epsilon$,
        $v_i = -A\omega_0 \sin\epsilon$.

Dividing each side of the second equation by the
corresponding side of the first equation and simplifying
completes the first part of the proof. If the initial
speed is zero, $\tan\epsilon = 0$ and $\epsilon = 0$ or whole number
multiples of π.

Again taking the equations for x_i and y_i, divide both sides
of the latter by ω_0, then square and add corresponding
sides of the two equations. Simplify to complete the
proof.

12.31   Since $mg = k(0.020$ m$)$, $k/m = g/(0.020$ m$)$ and
        $f = (1/2\pi)\sqrt{g/(0.020 \text{ m})} = 3.5$ Hz.

12.35   Energy is not conserved, but momentum is, so
        $m_b v_b = (m_b + M)v_0$.

Since $v_0 = 2\pi f x_0 = 8.48$ m/s,

$$(0.005\ 0 \text{ kg})v_b = (0.505 \text{ kg})(8.48 \text{ m/s})$$

and so $v_b = 0.86$ km/s.

12.47   Starting with $v = f\lambda$, multiply both sides by 2π to get $2\pi v = 2\pi f\lambda$, then divide by λ:

$$(2\pi/\lambda)v = 2\pi f = \omega.$$

12.55   Think of the wave as having been formed by raising a rectangular portion of water out, to form the trough, and up, to form the crest. That means an amount of work proportional to A, the height the water was raised. To both deepen the trough and raise the crest by A requires an additional amount of work proportional to $3A$, giving a net energy proportional to $4A$ in order to create such a wave.

12.57

12.61   From Eq. (12.10), $a_{max} = A(2\pi f)^2$.

$v = \lambda f$, so $f = v/\lambda = (2.0 \text{ m/s})/(1.0 \text{ m}) = 2.0 \text{ Hz}$;

hence $a_{max} = (0.040 \text{ m})(2\pi\, 2.0 \text{ s}^{-1})^2 = 6.3 \text{ m/s}^2$.

12.65   From Eq. (12.20),

$$v = \sqrt{F_T/\mu}$$
$$= \sqrt{(80 \text{ N})/(7.5 \times 10^{-3} \text{ kg/m})}$$
$$= 1.0 \times 10^2 \text{ m/s}.$$

12.71 $\Delta m = y(m/L)$.

At height y, $F_T = gy(m/L)$ hence $v = \sqrt{F_T/(m/L)} = \sqrt{gy}$. Thus the speed increases as y increases.

$V_{max} = \sqrt{gL}$. If a mass M is added, $F_T = gy(m/L) + Mg$, hence $v = \sqrt{F_T/(m/L)} = \sqrt{gy + MgL/m}$.

Answers to Discussion Questions

13.1   The outgoing pulse will trigger a positive peak at the
         microphone 29 ms after launch, which will hit the open end
         29 ms later, where it will be reflected with a 180° phase
         shift. Another 29 ms later the microphone will pick up a
         rarefaction and send out a negative peak. The scope will
         show a positive peak separated by 58 ms from a negative
         pulse.

13.3   The strings warm from the friction — they expand. There is
         a decrease in tension and a drop in speed and frequency.
         The wind instruments warm from the breath, increasing v
         and therefore f.

13.7   The triangular wave contains many strong overtones or
         Fourier components and so has a much greater high-
         frequency content, and sounds more "metallic." The sharper
         the bends in the string, the more high-frequency terms
         will be present. Plucking with the fingers, which are more
         rounded than a pick, will therefore generate fewer
         overtones and sound more mellow.

13.9   The answer lies in Young's Law of Strings (1800), which
         was discussed though not under its formal name. Remember
         that the point at which the string is struck must be an
         antinode because it is vibrating with maximum amplitude
         there. But the 7th harmonic has 7 antinodes and divides
         the string into 7 equal-length segments. Thus, 1/7th of
         the way down the string oscillating in the 7th harmonic,
         there must be a node. Hence, if we strike a string at that
         1/7th distance, we preclude the presence of the unpleasant
         7th harmonic.

13.13   The vibrating tuning fork transmits energy to the box,
         which is designed to have a standing wave pattern whose
         fundamental matches the frequency of the fork. The
         comparatively large box can impart energy to the
         surrounding air much more efficiently than could the fork,
         whose vibrations have a tiny amplitude. Sound comes out
         especially effectively from the open end of the box.
         Evidently, the lower the frequency, the larger the
         sounding box.

13.17   Stroking the rod sets it into standing-wave vibration with
         a node at the midpoint support. The attached piston

vibrates at that frequency and sets the air in the tube
vibrating. Adjusting the plunger allows a standing wave to
form in the doubly-closed tube and the powder is shaken
from the air-displacement antinodes toward the nodes,
where the air is still.

Answers to Multiple Choice Questions

1. a   3. d   5. e   7. d   9. d   11. b   13. c

15. a   17. b   19. a

Solutions to Problems

13.3 $v = (1500 \text{ Hz})\lambda$, so

$$\lambda = (0.50 \text{ m/s})/(1500 \text{ s}^{-1}) = 3.3 \times 10^4 \text{ m}.$$

13.7   The power is flowing uniformly through the sphere centered
on the source with radius 10 m, so

$$I = P/A = P/(4\pi R^2)$$
$$= (50 \text{ W})/(4\pi \times 100 \text{ m}^2) = 4.0 \times 10^{-2} \text{ W/m}^2.$$

The energy is given by

$$E = IAt$$
$$= (4.0 \times 10^{-2} \text{ W/m}^2)(1.0 \times 10^{-4} \text{ m}^2)(1.0 \text{ s})$$
$$= 4.0 \text{ MJ}.$$

13.17 $\Delta\beta = 10 \log_{10}1 = 0 \text{ dB}.$

13.21   With the previous problem in mind,

$$I = 10^{77/10}(10^{-12} \text{ W/m}^2) = 5.0 \times 10^{-5} \text{ W/m}^2.$$

13.33   Since intensity varies as the square of the pressure,

$$\beta = 10 \log_{10}(P/P_0)^2 = 20 \log_{10}(P/P_0).$$

13.41   The period of the beats is 0.99 s = $1/\Delta f$, so $\Delta f = 1.01$ Hz.

13.45 $\mu = (0.002\ 5\ \text{kg})/(1.00\ \text{m})$, so $f_1 = \dfrac{1}{2L}\sqrt{\dfrac{F_r}{\mu}} = 0.10$ kHz.

13.61   Since $v = f\lambda$ and λ is unchanged, the ratio of frequencies equals the ratio of speeds.

From Table 13.2, the ratio of speeds is
(331 m/s)/(970 m/s) = (600 Hz)/f;
$f = 1.76$ kHz.

13.73   From Eq. (13.17) $f_0 = (v + v_t) f_s/(v - v_t)$,
hence the beat frequency is $f_0 - f_s = 2 v_t f_s/(v - v_t)$.

Answers to Discussion Questions

14.1   Here are a few:

(a) The bore hole in the glass tube must be uniform.

(b) All of the mercury, including the stuff way up in the stem, must be at the same temperature.

(c) Generally, to speed up the thermometer's response the walls of the bulb are made thin, and that thinness makes it vulnerable to pressure variations which change its volume (via barometric changes, or hydrostatic pressure if it's immersed in a liquid).

(d) There is a variation in pressure in the mercury due to the different heights of the column.

(e) There is a difference in internal pressure depending on whether the thermometer is held vertically as opposed to horizontally.

(f) There are errors associated with the softness of the glass. If the thermometer is raised to a high temperature and then cooled rapidly, it might take weeks for the glass to return to its original volume. Try measuring the freezing point of water before and immediately after reading its boiling point — the difference can be as great as 1°C.

(g) The mere presence of the thermometer in a small system may change the temperature of the system.

(h) When measuring a changing temperature at any moment, the thermometer will always read warmer if the bath temperature is falling and colder if it is rising.

14.3   We know that the antimony expands on solidifying, as does water. Since that would be a very helpful trait for a casting material to have (it would fill all the fine details in the mold), it's reasonable to expect that is the answer to this question.

14.7   Being a poor conductor, the center of a boulder so treated
        would remain quite hot while the outside was cooled and
        contracted rapidly. Pressure would build up, there would
        be considerable internal stress, and the thing would
        rupture at any flaw or weak spot.

14.9   Bulbs are cheaper thinner and can tolerate the changes in
        temperature associated with ordinary operation, which are
        fairly gradual. They are made of inexpensive glass with a
        relatively large β, so a drop of water (or latex paint)
        can cause enough contraction and stress to shatter a hot
        light bulb. Clearly, outdoor lamps have to be protected
        from rain and snow. The heating in a flashbulb is so rapid
        even the thin walls tend to burst and they therefore are
        usually enclosed in a tough plastic film to keep them from
        shattering.

14.11   At a temperature of even –1°C it takes about 140 atm of
        pressure (≈ 2000 lb/in^2) to melt ice, so the problem was
        probably that the snow was just too cold.

14.13   The liquid will expand rapidly, its density decreasing as
        the density of the vapor increases. The surface meniscus
        will flatten out and then disappear altogether when the
        density of liquid and vapor are equal at a pressure of
        7.38 MPa.

14.15   Yes. They were once called permanent because they could
        not at first be liquefied, which suggests a weak
        intermolecular cohesive force. This, in turn, suggests
        ideal behavior.

14.17 $PV = nRT$, so both pressures must be equal since everything
        else is the same. The speeds of the hydrogen molecules
        must be greater than those of the nitrogen because the
        average KE is the same from Eq. (15.14). The pressures can
        be equal because the lighter hydrogens hit the walls at
        greater speeds and they do it more frequently because they
        traverse the chamber more quickly. The pressure is
        proportional to the average KE via Eq. (14.13).

14.19   Remember that the pressure in the room is more or less
         constant. Increasing T increases the average KE, which
         increases the net KE and the P but leads to an over-
         pressure and an outward current of air. The room leaks
         warm air to the outside. The temperature goes up because
         it's dependent on the average KE of each molecule, which
         is higher. The pressure remains the same because it
         depends on both the number of molecules per unit volume
         and their average KE. Compare Eqs.(14.13) and (14.14).

Answers to Multiple Choice Questions

1. b   3. a   5. c   7. d   9. a   11. b   13. a

15. a   17. b   19. e   21. a

Solutions to Problems

14.3 $0°F = -17.8°C$, $100°F = 37.8°C$,
         hence $\Delta T = 37.8 - (-17.8) = 55.6$ K.

14.11   Since aluminum has a linear expansion coefficient
         of 25×10^{-6} K^{-1}

$$\Delta L = (25 \times 10^{-6} \ K^{-1})(10 \ m)(20 \ K) = 5.0 \ mm.$$

14.19   A change of $180°F$ is a change of 100 K. The change in
         volume is given by

$$\Delta V = \beta V_0 \Delta T$$
$$= (182 \times 10^{-6} \ K^{-1})(0.50 \times 10^{-6} \ m^3)(100 \ K)$$
$$= 0.009 \ 1 \times 10^{-6} \ m^3.$$

The new volume is given by

$$V = 0.50 \ cm^3 + 0.009 \ 1 \ cm^3 = 0.51 \ cm^3.$$

14.27 The period is given by

$$\tau = 2\pi\sqrt{L/g} = 2\pi\sqrt{(1.000 \text{ m})/g} = 2.006 \text{ s}.$$

$$\Delta L = \alpha L_0 \Delta T = (25 \times 10^{-6} \text{ K}^{-1})(1.000 \text{ m})(-20 \text{ K}) = -0.50 \text{ mm}.$$

Then $L = 0.999\,5$ m, and again by $\tau = 2\pi\sqrt{L/g}$, $\tau = 2.005\,9$ s.

As for the clock, it ran fast.

14.35 $P_i V_i / T_i = P_f V_f / T_f$, so

$$(99 \text{ kPa})(1200 \text{ cm}^3)/(288 \text{ K}) = (101.3 \text{ kPa})V_f/(273 \text{ K}).$$

Then $V_f = 1.1 \times 10^3$ cm^3.

14.43 Each oxygen molecule has a mass of 32 u, hence 16 g is ½ mol and 16.0 kg is 500 mol.

$$(6.022 \times 10^{23} \text{ molecules/mol}) \times (500 \text{ mol})$$
$$= 3.01 \times 10^{26} \text{ molecules}.$$

14.51 From Eq. (14.15),

$$v_{rms} = \sqrt{3k_B T/m}$$
$$= [3(1.380\,66 \times 10^{-23} \text{ J/K})(293.15 \text{ K})/$$
$$(5.313\,6 \times 10^{-26} \text{ kg})]^{1/2}$$
$$= 478.03 \text{ m/s}.$$

14.63 Begin with Eq. (14.12), in the form $P = Nm(V^2)_{av}/3V$.

Nm is the total mass, so $Nm/V = \rho$, and hence

$$(3P/\rho)^{1/2} = \sqrt{(v^2)_{av}} = v_{rms}.$$

14.67   The nitrogen expands from 1.0 liter to 4.0 liters and the pressure drops to (1/4)2.0 atm = 0.5 atm.

The oxygen expands from 3.0 liters to 4.0 liters and its pressure ($P_i V_i = P_f V_f$) drops to (3/4)5.0 atm = 3.75 atm.

Hence the total pressure is

0.5 atm + 3.75 atm = 4.25 atm = 4.3 atm.

Answers to Discussion Questions

15.3   Bicycling consumes the same amount of energy per hour as
        shivering but converts 80% of it into thermal energy.
        Shivering, which doesn't perform any work to speak of,
        converts chemical energy almost completely into thermal
        energy. We shiver when the body loses too much heat and
        needs a quick infusion of thermal energy.

15.5   Copper is a much better conductor than iron or steel, but
        iron is cheaper and chemically a better material to cook
        food on. The copper layer is there to rapidly and
        uniformly distribute the heat by conduction. To the same
        end, good pots are generally made thick-bottomed, but
        that's not necessary if one just wants to boil water.

15.7 Q/t should be proportional to the exposed area A, the
        temperature difference between the body and the fluid ΔT,
        and the physical parameters that describe the particular
        situation, such as geometry, orientation, wind speed, etc.
        All of the latter are combined into a constant of
        proportionality, K_c, called the convection coefficient,
        which is determined experimentally. Hence, $Q/t = K_c A \Delta T$.
        For a human on a windless day, $K_c \approx 5$ kcal/m$^2 \cdot$ h $\cdot$ C$^\circ$, and A is
        the exposed area of the skin.

15.9   We tend to be more comfortable in a dry-air environment
        when it's hot because the body can perspire and control
        its temperature more effectively when evaporation is
        rapid. The rate of evaporation decreases as the amount of
        water vapor in the air increases. When the body gets too
        warm, the blood vessels at the skin dilate so as to bring
        more blood to the surface, which makes the skin red.

15.11   The ice forms first where it is in contact with the tray
        (which is a better conductor than the air) and then across
        the top, thereby encapsulating some water. Since ice is
        not a very good conductor, the remaining water takes a
        while to freeze. The warm water will be cooled more
        effectively, especially by evaporation, and enough of it
        may evaporate so that it will win the race. Even so, there
        are a lot of variables and the cool water often freezes
        first.

15.13   The red arrows indicate the rate of flow of heat that
        decreases in the exposed rod where there are losses

(mostly via convection), and remains constant in the insulated rod where there are not. In the exposed rod, there is less and less thermal energy available and the flow diminishes. Correspondingly, the slope of the *T-d* curve, which is the *temperature gradient*, also diminishes. In the insulated rod, the heat flow is uniform and the temperature gradient constant; that is, the slope is constant. Heat is transported by the temperature gradient much as a liquid is propelled through a pipe by a potential energy gradient. If heat were an incompressible liquid, it would flow as in part (b).

15.17   Wearing the hair inside will trap air that will not be disturbed by external winds. The low thermal conductivity of the air layer is what is important and that is better achieved with the fur inside. To stay cool, wear light-colored, porous, light-weight clothes. These will reflect radiant energy and facilitate evaporation of sweat. If there isn't much water to be had, evaporation of perspiration should be restrained as much as possible.

15.19   Thermal radiation from the Sun will strike a portion of your suit and it will absorb energy at a rate dependent upon its surrounding face characteristics. The intensity of that incident radiation is determined in part by the distance to the source — the farther away, the less power per unit area. Moreover, the suit will radiate over its entire surface at a rate dependent on its temperature (which, in turn, is partly determined by the body's output). The same sort of thing will happen to the thermometer. It will read an equilibrium temperature that is determined, among other things, by the fraction of its area illuminated. At the distance of the Earth from the Sun, a sphere in thermal equilibrium will stabilize at about 290 K, not far from room temperature. Thus, the thermometer reads its own temperature and not that of space.

Answers to Multiple Choice Questions

1. e   3. c   5. c   7. b   9. a   11. c   13. a

15. b   17. b   19. c

Solutions to Problems

15.7   From Eq. (15.1),

$Q = cm\Delta T = (1.00 \text{ kcal/kg} \cdot \text{K})(30 \text{ kg})\Delta T = 500 \text{ kcal}$,

so $\Delta T = 17$ K.

15.15   The heat lost by the iron equals the heat gained by the water:

$(100 \text{ g})(0.113)(1.00 \text{ cal/g} \cdot \text{K})(50 \text{ K}) = m(1.00 \text{ cal/g} \cdot \text{K})(5 \text{ K})$.

So $m = 113$ g.

15.23   The kinetic energy of the bullet is converted into thermal energy which, going partly into the lead bullet and partly into the wood block, warms the block and the bullet trapped in it by the same number of degrees.

540 ft·lb = 732 J, so

$$Q = 732 \text{ J} = Q_L + Q_W = c_L m_L \Delta T + c_W m_W \Delta T$$
$$= \Delta T[(130 \text{ J/kg} \cdot \text{K})(10.24 \times 10^{-3} \text{ kg}) +$$
$$(1700 \text{ J/kg} \cdot \text{K})(1.0 \text{ kg})]$$
$$= \Delta T(1701 \text{ J/K}).$$

$\Delta T = +0.43$ K.

15.27   The heat went partly into the aluminum and partly into the water, so

$$Q = Q_A + Q_W$$
$$= (900 \text{ J/kg} \cdot \text{K})(0.50 \text{ kg} + 0.60 \text{ kg})(12 \text{ K})$$
$$+ (4186 \text{ J/kg} \cdot \text{k})(2.00 \text{ kg})(12 \text{ K})$$
$$= 112 \ 344 \text{ J} = 1.1 \times 10^5 \text{ J}$$
$$= 27 \text{ kcal.}$$

15.35   This requires an input of

$$Q = cm\Delta T = (4186 \text{ J/kg} \cdot \text{K})(1.00 \text{ kg})(100 \text{ K}) = 418 \ 600 \text{ J.}$$

The heat of combustion of hard coal is 33 MJ/kg, hence we need

$$(0.418 \ 6 \text{ MJ})/(33 \text{ MJ/kg}) = 0.012 \ 7 \text{ kg} = 12.7 \text{ g.}$$

15.67   The amount of heat needed to raise the water from 0°C to 80°C is the same as the amount needed to melt the ice at 0°C regardless of how much water and ice there is, provided they're equal. Hence the ice melts and we end up with twice as much water at 0°C.

15.71   The thermal energy is given by

$$Q = c_w m_w \Delta T_w = (4186 \text{ J/kg} \cdot \text{K})(0.900 \text{ kg})(95 \text{ K})$$
$$= 0.357 \ 90 \text{ MJ} = 0.36 \text{ MJ.}$$

Added to the silver,

$$0.36 \text{ MJ} = c_s m_s \Delta T_s = (230 \text{ J/kg·K})(1.20 \text{ kg})(\Delta T_s).$$

Then $\Delta T_s = 1296.7$ K and $T_f = 1575$ K. But the melting point of silver is only about 1234 K, so it cannot get up to 1575 K without liquefying.

15.71   (Cont'd)

When the melting point is reached, the heat used up is given by

Q = (230 J/kg·K)(1.20 kg)(960.8°C − 5.0°C) = 0.268 30 MJ,

leaving 0.094 099 MJ.

Melting 1 kg of silver requires

Q = mL_f = (1.00 kg)(109 × 10³ J/kg) = 109 kJ.

So not all the silver melts, and T_f = 1233.95 K.

Answers to Discussion Questions

16.1   Air is drawn from A to B as the piston moves out. From
         B to C, it is compressed adiabatically and rises in
         temperature. Fuel is sprayed in at C and immediately
         explodes because of the high temperature. C–D is the
         isobaric combustion leg when heat enters. At D, the fuel
         is burnt and the piston continues to move outward. D–E is
         the rapid and therefore adiabatic expansion where the hot
         gas does work. E is the end of the cycle and it's there
         that the exhaust valve opens. E–B is an isovolumic drop to
         the outside pressure, accompanied by the loss of heat via
         the exhausting of hot gas. With the exhaust valve still
         open, the remaining gas is ejected as the piston moves
         inward from B to A.

16.3   Steam enters via the right-hand valve, behind the piston,
         and drives it to the left as the slide valve moves to the
         right. Steam in the left half of the cylinder escapes
         through the left-hand valve to the exhaust port and from
         there to the low-pressure condenser. The right-hand valve
         then closes, the steam expands, and the piston moves left.
         Now right-hand valve opens to the exhaust, steam enters
         via the left-hand valve and the piston moves right.

16.5   Warm, moist low-density air rises (via thermals) doing
         work on the atmosphere as it expands. Accordingly, since
         the process occurs quickly, and gases are very poor
         conductors, it will be essentially adiabatic. The internal
         energy of the uprush of air will decrease and its
         temperature will drop. Water will then condense out,
         forming a cloud.

16.7   The Sun provides the energy stored chemically in wood,
         coal, and oil. Order increases locally as the plants and
         animals storing solar energy grow, ultimately to form
         fossil fuels, but increase in the disorder of the Sun
         overbalances that.

16.9   The Sun evaporates water that comes down to the rivers and
         reservoirs as rain, thereafter to pour down a waterfall
         and drive a turbine. In effect, the Sun does work against
         gravity.

16.11   The process is adiabatic and so isentropic. Work is done
         by the gas on the piston. Therefore, the internal energy

decreases, and the temperature (T) drops. The volume obviously doubles and the pressure decreases. The entropy must stay constant since $Q = 0$.

16.13   The engine actually exhausts gas at a much higher temperature than 300 K. Fuel is incompletely burned, heat is conducted and radiated from the engine, and friction is present as well.

16.15   Yes, it is possible, but with so many molecules the likelihood of all of them being in one corner is minute. Of course, if there were only 1 molecule flying around, the chance of "all the molecules" being in one corner would be fairly high. It's less likely with 10 molecules and much less still with 10^{28}.

Answers to Multiple Choice Questions

1. b   3. d   5. d   7. b   9. e   11. b   13. c

15. a   17. d   19. b

Solutions to Problems

16.7   100 cal = 418.6 J. From Eq. (16.2),

$$\Delta U = Q - W = 418.6 \text{ J} - (-100.4 \text{ J}) = 519 \text{ J}.$$

16.15   98.6°F = 310 K. Then from Eq. (16.11),

$$e_c = 1 - (T_L/T_H) = 1 - (293 \text{ K})/(310 \text{ K}) = 5.5\%.$$

16.23

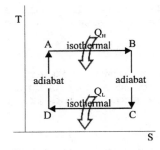

16.31   From Eq.(16.2), $\Delta U = Q - W$. $Q = 0$, so

$-\Delta U = W = P\Delta V = (99 \text{ kPa}) \cdot (4950 \times 10^{-6} \text{ m}^3) = 490 \text{ J}.$

The removal of that much heat corresponds to a condensation determined by the equation $Q = mL_v$.

Solving for m yields

$m = (490 \text{ J})/(2259 \text{ kJ/kg}) = 0.22 \text{ g}.$

16.35 $T_1 = 273$ K.

(a) From the diagram the lowest pressure is P_2. From Eq.(16.7), $P_2 = P_1(V_1/V_2)^\gamma$.

Since hydrogen is diatomic, $\gamma = 1.4$ and $P_2 = (1 \text{ atm})(\frac{1}{2})^{1.4} = 0.379 \text{ atm} = 0.0384 \text{ MPa}.$

(b) From Eq.(16.8), $T_2 = P_2V_2T_1/P_1V_1 = 206.9 \text{ K} = 207 \text{ K}.$

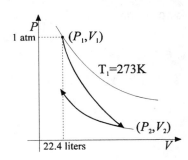

16.43   2000 lb is equivalent to 907.2 kg. The amount of heat
        needed to vaporize that much water is given by

$$Q_L = mL_f = (907.2 \text{ kg})(333.7 \text{ kJ/kg}) = 302.7 \text{ MJ}.$$

From Eqs.(16.13) and (16.15),

$Q_L/W_i = T_L/(T_H - T_L)$, so

$$W_i = Q_L(T_H - T_L)/T_L$$
$$= (302.7 \text{ MJ}) \cdot (31.1 \text{ K})/(273 \text{ K})$$
$$= 34.5 \text{ MJ (per day), or } 399 \text{ W.}$$

16.47 $Q_L/W_i = \eta = T_L/(T_H - T_L)$.

$$W_i(\text{per minute}) = (360 \text{ J/min})(31 \text{ K}/269 \text{ K})$$
$$= 41.49 \text{ J/min} = 0.69 \text{ W.}$$

16.51   From Eq.(16.10), $e = 1 - 450/1200 = 62.5\%$.

As an engine, it has $Q_H = 1200$ J, $Q_L = 450$ J, and
$W = 750$ J, and those numbers must be the same if the
engine is reversible even if now 1200 J is heat-out rather
than heat-in, etc.

But here work-in is 1800 J and not 750 J, so the engine is
not reversible.

Answers to Discussion Questions

17.1   The paper is electrically polarized by the field, and is
        therefore attracted to the charged conductor (or sheet of
        glass). It is only after a little while that charge is
        transferred to it from the highly charged object. At that
        point, the paper has a net charge with the same polarity
        as the conductor (or glass) and is repelled.

17.3   The field is strongest in
        the region between the
        charges. Because the walls
        are conducting, the field
        lines are perpendicular
        to the walls and because
        the chamber is grounded
        there is no outer-surface
        charge and no field
        outside.

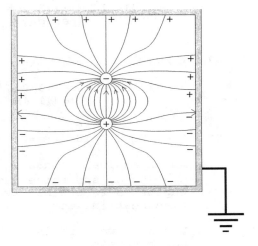

17.7   Rubbing the tube electrified it and the free charges
        flowed along the moistened string, which was a fairly good
        conductor. The lead ball became charged as if it had been
        touched directly by the glass rod. It then polarized the
        leaf, which was attracted upward towards it.

17.9   As we will discuss later, the typewriter radiates an
        electromagnetic wave that the antenna wire picks up and
        superimposes on the signal coming down from the roof. One
        thing to do is to wrap the antenna wire, in the vicinity
        of the typewriter, with aluminum foil and ground it, thus
        shielding it. Another option (though we will not discuss
        it until later), is to reorient the portion of the wire
        near the typewriter so it is not along the radiated E-
        field.

17.15   *The case of attraction is always ambiguous* — the elephant
        might be either negative or neutral, whereupon the
        positive ball would induce a negative charge and be
        attracted. The repulsion unambiguously means that the
        elephant was negative.

Answers to Multiple Choice Questions

1. c   3. b   5. c   7. a   9. a   11. d   13. b

15. c   17. d   19. e

Solutions to Problems

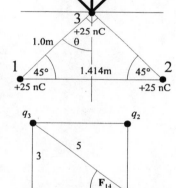

17.11   By symmetry, the net force is $2(F_{31} \cos \theta) = 2(F_{32} \cos \theta)$, since the horizontal components cancel; hence $F = 2(kqq/r^2) \cos \theta$; $\theta = 45°$ and by the Pythagorean Theorem $r = 1$, so $F = 2 \cos \theta \times (9.0 \times 10^9 \text{ N·m}^2/\text{C}^2)(+25 \times 10^{-9} \text{ C})^2/ (1.0 \text{ m})^2 = 8.0 \ \mu\text{N}$, upward, along the perpendicular bisector of the hypotenuse.

17.19 $F_{12} = kq_1q_2/r_{12}^2 = 3.6$ N;

$F_{13} = kq_1q_3/r_{13}^2 = 4.5$ N;

$F_{14} = kq_1q_4/r_{14}^2 = -1.8$ N;

$F_{x1} = (-1.8 \text{ N}) + (4.5 \text{ N})4/5 = 1.8$ N; $F_{y1} = -(3.6 \text{ N}) - (4.5 \text{ N})3/5 = -6.3$ N. The net force, then, is down and to the right in the figure, with its magnitude given by $F_1 = \sqrt{1.8^2 + 6.3^2}$ N = 6.6 N; the direction of the net force is below the horizontal by an angle of $\theta = \tan^{-1}(6.3/1.8) = 74°$.

17.33   **E** at the third vertex has a net component only along a line bisecting that angle (and pointing away from the triangle). Total $E = 2(kq/r^2) \cos 30° = 2(9.0 \times 10^9 \text{ N·m}^2/\text{C}^2)(20 \times 10^{-6} \text{ C})(0.866)/(2.0 \text{ m})^2$ and $E = 7.8 \times 10^4$ N/C.

17.39   They should be charged with $+q$ on the bottom plate and $-q$ on the top where $E = \sigma/\epsilon_0 = q/A\epsilon_0$; then $q = (1000 \text{ N/C})(1.00 \text{ m} \times 0.50 \text{ m})(8.85 \times 10^{-12} \text{ C}^2/\text{N·m}^2) = 4.4$ nC.

17.47   The force acting on the electron is $q_eE = m_ea$. Then $a = q_eE/m_e = (1.60 \times 10^{-19} \text{ C})(1.5 \times 10^4 \text{ N/C})/(9.11 \times 10^{-31} \text{ kg}) = 2.6 \times 10^{15}$ m/s^2.

17.51   Draw a spherical Gaussian surface inside the ball at a
         radius r from the center. By symmetry the E-field must be
         radial. $\rho = Q/V = Q/(4\pi R^3/3)$; the charge inside the
         surface is $\rho(4\pi r^3/3)$, hence $E(4\pi r^2) = \rho(4\pi r^3/3)/\epsilon_0$; and $E = r\rho/3\epsilon_0 = rQ/4\pi\epsilon_0 R^3$. Inside the charge distribution, then,
         the field increases linearly with distance from the
         center.

Answers to Discussion Questions

18.3   The neutral conductor becomes polarized with a negative induced charge near the original sphere and a positive charge on the far side. These charges, in turn, contribute to the potential, with the negative charges dominating and reducing the potential at the original sphere. Accordingly, the potential of the original sphere (that is, inside) is lower. Outside, it drops rapidly to a finite value that is constant across the neutral conductor.

18.5   Since no charge flows from or to the neutral conductor while touching the inside wall, the two must be at the same potential, which is uniform in the cavity, implying that the conductor assumes the potential of the field-free region. In a region where there is a field, the conductor will distort the field pattern and change the potential. If more charge is added to the outer conductor, the potential inside increases and the potential of the neutral body increases.

18.9   The field lines go from the positive charges to the negative ones. There is no field at the very center. The equipotentials surround each of the charges, being negative in the vicinity of the negative charge. There are two zero equipotential planes that pass through the center point.

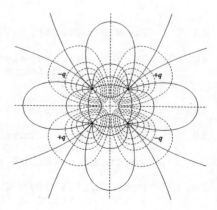

18.11   Bringing a positive charge to a point in space introduces an outwardly directed field and a positive potential everywhere in the region; a negative charge does just the opposite, contributing a negative potential and so lowering the net potential everywhere. Imagine a gold-leaf electroscope attached to a positively charged plate. The electroscope reads the potential of the plate. Now bring a positive charge near the plate; positive charge from the plate will be repelled back to the scope, the leaves will part even more, and it will thereby indicate an increase in potential.

18.15   There cannot be a charge since that would produce a field and therefore a potential gradient, but the potential is constant.

18.19   The electric field lines are perpendicular to the grid of curved equipotentials. The field lines, which become fairly straight, converge to point P (from the anode on the right) and then diverge away from P (on the left). Electrons in the beam approaching from the left are consequently accelerated toward P.

Answers to Multiple Choice Questions

  1. b   3. c   5. b   7. d   9. d   11. d   13. c

15. d   17. b   19. b

Solutions to Problems

18.3   From Eq. (18.6), $\Delta V = \pm Ed = \pm(1.00 \text{ V/m})(0.10 \text{ m}) = \pm 0.10 \text{ V}$.

18.17   To suspend the electron, the upward force $q_e E$ must equal the downward force $m_e g$. So $q_e E = q_e \Delta V/d = m_e g$; $\Delta V = m_e g d/q_e = 5.58 \times 10^{-12} \text{ V}$.

18.21 $Ed = \Delta V$, so $E = \Delta V/d = (85 \text{ mV})/(8 \text{ nm}) = 11 \text{ MV/m}$.

18.27   The voltage across the capacitor is so high that it can be quite dangerous. $Q = CV = (500 \text{ pF})(20 \text{ kV}) = 1 \times 10^{-5} \text{ C}$, which is a lot of charge.

18.33   From Eq. (18.12), $C = 10\epsilon_0 A/d = 10(8.85 \times 10^{-12} \text{ C}^2/\text{N} \cdot \text{m}^2)(100 \times 10^{-4} \text{ m}^2)/(1.0 \times 10^{-3} \text{ m}) = 8.9 \times 10^{-10} \text{ F}$.

18.39   The stack may be thought of as 100 capacitors. They are all in parallel and act like a single capacitor with an area equal to the total area of the paper; $C = 100(4.1\epsilon_0)A/d = 0.99$ μF.

18.51   They are all in parallel, hence $C = 9$ pF.

18.59   From Problem 18.55 we have that $C = 30$ pF, hence from Eq. (18.16) $PE_E = \frac{1}{2}CV^2 = \frac{1}{2}(30$ pF$)(12$ V$)^2 = 2.2$ nJ.

18.61   The system is actually two capacitors in parallel with the center sheet serving as a plate for both (recall Prob. 18.39). For each capacitor $C = \epsilon A/d = (4.5)(8.85 \times 10^{-12}$ C^2/N$\cdot$m$^2)(0.30$ m$^2)/(1.0 \times 10^{-3}$ m$) = 12$ nF, and so the net capacitance is 24 nF.

Answers to Discussion Questions

19.1   The flow of heat along a metal rod is proportional to the temperature difference across its ends, just as the flow of charge is proportional to the voltage difference. In metals, free electrons serve to transport both electrical and thermal currents.

19.5   This arrangement is for controlling a light bulb from two locations. Either switch will turn the bulb on or off independent of the other. As shown in Fig. Q5, the lamp is on. Throwing either switch open-circuits the bulb, shutting it off (see figure).

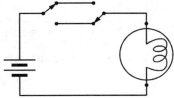

19.7   The gasoline engine powers a separate electrical generator that operates all systems and sends its excess current to the battery to recharge it. Once the engine is running, the battery can be completely removed from the circuit. Leaving off all unnecessary electrical devices will allow more current to be provided to recharge the battery.

19.11 The ammonium chloride dissociates into NH^+ (which, while current is circulating, migrates to the carbon rod) and Cl^-. Chlorine ions combine with the zinc electrode to make zinc chloride. If it were really dry, there would be no migration of the ions and it wouldn't work. Until only a few decades ago, the cell's outer casing was the thin zinc can, simply wrapped in a cardboard insulating sleeve. More often than not, the zinc would be eaten away and the ammonium chloride paste would leak out and corrode your flashlight. Even with today's steel casings, it's still not wise to store electronic equipment with batteries inside them for long periods.

19.15 With current fanning out radially from the point of impact, there will be a voltage drop across the animal, and if its resistance is not very much greater than the ground's, a sizeable current will pass through it. Squatting is better than standing because it makes one a less likely target for a direct hit by lightning, and squatting does not, like lying down, expose one to the risk of electrocution from radial current.

Answers to Multiple Choice Questions

1. c   3. d   5. c   7. e   9. b   11. b   13. a

15. c   17. d   19. c   21. a

Solutions to Problems

19.5 $I = \Delta Q/\Delta t$; so $\Delta Q = (180 \text{ A})(2.0 \text{ s}) = 3.6 \times 10^2$ C.

19.17   (a) A and B for the horn, (b) A and E for the headlights, (c) A, D and C for the tail lights, (d) A and D for the parking lights, (e) A and F for the dome light. The side marker lights go on when the headlights are on.

19.23 ΔQ is the area under the I versus t curve which is $\frac{1}{2}(7.0 \text{ A} - 3.0 \text{ A})(6.0 \text{ h} \times 60 \text{ min/h} \times 60 \text{ s/min}) + (3.0 \text{ A})(21\,600 \text{ s}) = 43\,200 \text{ C} + 64\,800 \text{ C} = 0.11$ MC.

19.27   From Eq.(19.5), $V = IR$; so $R = V/I = (100 \text{ kV})/(2.0 \text{ }\mu\text{A}) = 50 \times 10^9 \text{ }\Omega$.

19.35   From Eq.(19.6) and Table 19.2, $A = \rho L/R = (1.7 \times 10^{-8} \text{ }\Omega \cdot \text{m})(1609.3 \text{ m})/(10 \text{ }\Omega) = 2.736 \times 10^{-6} \text{ m}^2$; since for a circular cross-section $A = \pi r^2$, $r^2 = 8.71 \times 10^{-7} \text{ m}^2$; the diameter, $2r$, is 1.9 mm.

19.45   From Eq.(19.9), $P = IV = (180 \text{ A})(12 \text{ V}) = 2.2$ kW.

19.57   Current per strand $I_s = 0.012$ A; $V_s = I_sR = 2.4 \times 10^{-5}$ V; so (b) across the wire $V = 2.4 \times 10^{-5}$ V; (a) $R = V/I = (2.4 \times 10^{-5} \text{ V})/(0.12 \text{ A}) = 2.0 \times 10^{-4} \text{ }\Omega$; (c) From Eq.(19.10), $I_s^2 R_s = 2.9 \times 10^{-7}$ W.

19.61 $P = IV = [(10 \text{ }\mu\text{C})/(1 \text{ s})](3 \text{ MV}) = 30$ W, neglecting friction losses.

19.67 $\Delta R/R_0 = \alpha_0 \Delta T = (0.004\,5 \text{ K}^{-1})(600 \text{ K}) = 2.7$; clearly α_0 is not constant and the calculation is quite poor. At 2500° C α_0 can be crudely approximated graphically to be about 0.02 K^{-1}. For this α_0, $\Delta R/R_0 = \alpha_0 \Delta T \approx (0.02 \text{ K}^{-1})(600 \text{ K}) \approx 12$.

Answers to Discussion Questions

20.3 This circuit was supposed to simply illustrate the Node Rule at G; 12 amps flow in, 12 amps flow out. But they overlooked something. A-H-G and B-H-G are short circuits; A and G, and B and G are all at the same potential. That means that no current flows through the resistors in branches A-G and B-G, and the diagram is incorrect as labeled.

20.5 The switch is closed and the capacitor charges up slowly (depending on RC). As it does, the voltage across it increases until the breakdown voltage of the lamp is reached, at which point the capacitor rapidly discharges through the lamp, which flashes. The voltage across the capacitor, and therefore the lamp as well, drops rapidly only to slowly build up again for a repeat performance.

20.7 The circuit first sees the edge of the pulse and a voltage of V. Charge and voltage both gradually build on C until the signal drops to 0 at which point they both decay. If the time constant is short enough, V_C will reach nearly 0 before the next pulse arrives.

Since the drop across the capacitor is initially zero, V_R starts at a maximum and decays exponentially to 0. When the input signal drops to 0, it's as if the input terminals were shorted. The voltage in the circuit comes from the charged capacitor, and the resistor is across its terminals. When the upper input circuit terminal was positive, the current circulated clockwise. When the input is shorted, current (discharged from the capacitor) circulates counterclockwise. Hence, the voltage polarity on the resistor is reversed and decays from there to zero as the capacitor discharges. In general, $V_R = V - V_C$.

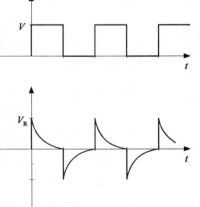

20.9 The benchmark of *direct current* is that it does not change direction, which means that a dc voltage must not change polarity or sign. It certainly can vary — it need not be constant — but it cannot reverse itself. a, c, d, e, g, and h are all dc. b and f are known as alternating current, ac.

20.11   1 is brightest; 2 and 3 are equally bright, though less so than 1; no current passes through 4, 5, or 6, and so they are off altogether.

20.13   The resistance of the middle branch is half that of the upper branch; it draws twice the current and so 3 is brighter than 1 or 2, which have the same current and are equally dim.

20.15 R_1 and R_5 have the most current; R_2, R_3 and R_4 have less; R_6 and R_7 have none. (R_7 has no voltage across it and there is no closed circuit through R_6).

Answers to Multiple Choice Questions

1. d   3. d   5. c   7. c   9. a   11. b   13. d

15. c   17. a   19. b   21. b

Solutions to Problems

20.1 $R_e = 20\ \Omega$, $I = V/R_e = 0.60$ A before; after the wire is attached, the current in the left side (and through the wire) is given by $I = V/R_e = (12\ \text{V})/(8.0\ \Omega) = 1.5$ A. The 12-Ω bulb is short-circuited and goes out.

20.13   2 Ω, 3 Ω, and 6 Ω in parallel equals 1 Ω, in parallel with 1 Ω equals ½ Ω, in series with 5 Ω is 5.5 Ω.

20.23 $P = IV$, 80 W $= I(20\ \text{V})$, so $I = 4.0$ A; $V = IR$ where V is the voltage across the resistor, so $R = (60\ \text{V} - 20\ \text{V})/(4.0\ \text{A}) = 10\ \Omega$.

20.27   Charge is 63% of maximum at $t = RC = (5.0\ \text{k}\Omega)(800\ \mu\text{F}) = 4.0$ s.

20.33   9 A through 17 Ω;
9 A splits, 1 part out of
10 or 0.1(9 A) = 0.9 A
through 9 Ω, and
(9/10)(9 A) = 8.1 A
through 1 Ω;
9 A enters and
leaves the
next node;
3 Ω + 1 Ω =
4 Ω, and
4 Ω + 1 Ω = 5 Ω,
so the next portion consists
of 4 Ω in parallel with 2 Ω in
parallel with 5 Ω yields a
resistance for this portion of
1.05 Ω. Since the total current
through this portion must be 9 A, V =
(9 A)(1.05 Ω) = 9.47 V hence current
through the 3 Ω and 1 Ω resistors is
(9.47 V)/(4 Ω) = 2.4 A or 2 A. Through the
2 Ω passes 4.7 A or 5 A. Through the 4 Ω and 1 Ω
resistors passes 1.9 A or 2 A.

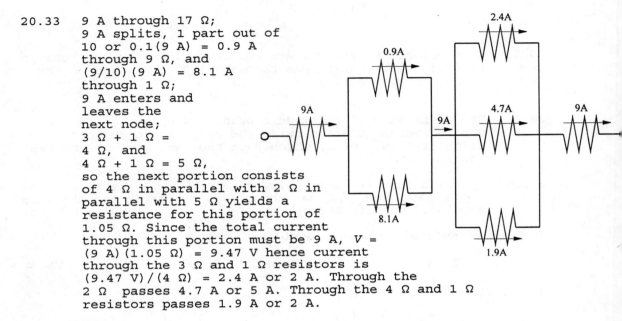

20.49   The voltage across *A-D-C* is 12 V hence the current is
(12 V)/(4 Ω) = 3 A. The voltage across the 2-Ω resistor in
loop *A-B-C-A* is 12 V – 6 V hence the current through it is
3 A, and from the Node Rule the current through the 12-V
battery is 6 A.

20.61   The voltage between the top and bottom nodes is
48 V – (3 A)(4 Ω) = +36 V, which equals R_1(1.0 A) - 10 A.
Hence, R_1 = 46 Ω. +36 V also equals (25 Ω)(2.0 A) - $\mathscr{E}_1$, so
$\mathscr{E}_1$ = 14 V.

20.67 $I_2 = I_6$ = 2.0 A, $I_4 = I_5$ = 5.0 A, V = (10 Ω + 5.0 Ω)(2 A) =
30 V = (2.0 Ω + R)(5.0 A); R = 4.0 Ω. I_3 = 1.0 A (which
was given but can be derived by I_3 = (30 V)/(30 Ω)), and I_1
= $I_2 + I_3 + I_4$ = 8.0 A.

Answers to Discussion Questions

21.1   The electrostatic E-field of a
        point charge should have
        spherical symmetry because
        the charge presumably has
        spherical symmetry and space
        is isotropic. Once the charge
        is set moving, the axis of that
        motion represents a
        distinguishable direction and
        both the E and B fields will now
        only be axially symmetric (one
        circular, one radial).

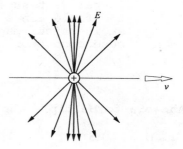

21.3   Each north pole would experience a torque of $q_m B_N R_N$ and
        each south pole a torque of $q_m B_S R_S$ in the opposite
        direction. These must cancel since there's no rotation,
        hence they are equal and $B_S/B_N = R_N/R_S$; the field varies
        inversely with the radial distance.

21.7   The field of the coil rapidly changes magnitude and
        direction and that tends to mix up the domain structure.
        Since $B \to 0$ as the coil is removed, the domains cannot
        continue to realign themselves with the field of the coil.

21.11   Far from any long wiggly wire the field must resemble that
        of a straight wire. Certainly if the bends are perfectly
        regular (e.g., an equilateral sawtooth pattern or a
        sinusoid) we could expect the same field (some distance
        away) as from a straight current. Since field lines do not
        cross we can imagine sets of circular lines around each
        tiny segment of the bent wire. These can be thought of as
        flattening out into planes perpendicular to the axis of
        the wire at distances large compared to the wiggles —
        remember the E-field of two equal charges. Ampère's Law
        makes no distinction between wiggly currents and straight
        ones. Experimentally, it is found that the setup will be
        unaffected by the introduction of a B-field. A solenoid is
        a spiralling wire, and if it has a projected length (end-
        to-end), it will have the external field of a straight
        wire.

21.15   Looking down on the left side, there is a counter-
        clockwise B-field and the north pole rotates
        counterclockwise around with the field. On the right,
        there is a downward current in the B-field of the magnet,
        and therefore the hanging rod experiences a perpendicular
        force that causes it to rotate clockwise.

21.17   The magnet's field was expelled from the disk when the
        latter went superconducting. The result is very much as if
        the field lines were bent up and away by the presence of
        an identical magnet (with like poles) within the
        superconductor that repels the first magnet.

Answers to Multiple Choice Questions

1. d   3. c   5. c   7. b   9. b   11. a   13. a

15. b   17. c   19. c   21. a

Solutions to Problems

21.1   From Eq.(21.7) 1 T = 1 N/(C·m/s); then
        1 T·m/A = 1 (N·s/C·m)·m/A = 1 (N/A)/A = 1 N/A^2.

21.13   From Eq.(21.5), $B_z \approx \mu_0 nI$;
        5.0×10^{-4} T = $(4\pi \times 10^{-7}$ T·m/A)[200/(0.10 m)]I;
        I = 0.20 A.

21.17   Since the charge on one electron is about 1.6×10^{-19} C,
        I = $(6.0 \times 10^{12}$ s$^{-1})(1.6 \times 10^{-19}$ C) = 9.61×10^{-7} A;
        From Eq.(21.2), then, $B = \mu_0 I/2\pi r = 1.3 \times 10^{-11}$ T, clockwise
        looking toward the source.

21.25   At z = 0 the equation reduces to Eq.(21.3). As z gets very
        large ($z \gg R$), R is negligible in the numerator and the
        equation becomes

$$B_z = \frac{\mu_0 IR^2}{2z^3}$$

21.31   Using Ampère's Law for a circular path in the plane of the
        torus and lying within the central hole ($r < R_i$), the
        field must be circular if it exists at all. But no current
        is encompassed, so $B\sum \Delta l = \mu \sum I$ = 0 and B = 0. The same
        would be true outside the torus where $r > R_0$. There the
        same number of wires carry current into the plane of the
        Ampèrean path loop as out of that plane. Hence $\sum I$ = 0 and
        B = 0. (This argument ignores the very small field whose
        lines pass through the torus perpendicular to its plane.
        This field is due to the net transport of charge around
        the torus — recall the discussion of a solenoid on p. 740;
        the field could be eliminated by winding the torus in an
        even number of layers.)

21.35   From Problem 21.33, overlapping the two fields from the sheets, it's clear that $B = 0$ outside and $B = 2(\frac{1}{2}\mu_0 i)$ inside. Alternatively, using Ampère's Law with a rectangular loop of width l with one side between the sheets, one above and the other two perpendicular to the sheets: $\sum B_{\parallel} \Delta l = \mu_0 \sum I$; $Bl + 0 + 0 + 0 + 0 = \mu_0 il$; and $B = \mu_0 i$, here B is perpendicular to two sides and zero over the third.

21.47   From Eq. (21.8), $\mathbf{F} = q\mathbf{v} \times \mathbf{B}$ and by the right-hand rule, the force is in the $-y$ direction.

21.67   From Eq. (21.13), $F_M/l = \mu_0 I_1 I_2 / 2\pi d = (4\pi \times 10^{-7}\ \text{T·m/A}) \times (10\ \text{A})^2 / (2\pi \times 1.0\ \text{m}) = 2.0 \times 10^{-5}\ \text{N}$; because the currents are antiparallel, the force is repulsive.

Answers to Discussion Questions

22.3   The handle of the device slips into a coil hidden in the base of the holder. Another coil is inside the handle and the two are in close proximity. Connecting the coil in the well to AC generates a time-varying B-field that passes through the plastic skin of the handle and produces a rapidly varying (60 Hz) induced emf and an induced current. That current is converted to DC and used to charge a battery in the handle, which powers the motor in the device. No charge and so no "electricity" passes from the base to the handle, although electrical energy certainly is transferred. The same scheme will work across human skin to power a mechanical heart.

22.5   As the magnet approaches, its B-field attempts to penetrate the ring. A supercurrent is thereby induced opposing the buildup of flux and the motion of the magnet. If the ring is free to move, it will lift off its support and hover (with its induced north pole down) above the north pole of the magnet. Finite, persistent supercurrents *on the surface* of the superconductor circulate in such a way as to shield the interior from the field. No flux enters the body of the superconductor; there is no change in flux, no E-field induced, and no bulk current.

22.7   To crank the generator without a load and therefore without a current (in the steady state), one need only overcome friction. By contrast, lighting the 100-W bulb requires twice the power used in lighting the 50-W bulb and that power must be supplied by the person turning the crank. Most people will find it difficult to keep the 100-W bulb lit very long.

22.9   Current would be induced in the loop and power transferred from the line to it — the time-varying B-field would induce an E-field. If a wire loop was in that region, the E-field would do work on the free electrons, transferring energy to them. The power company could detect a loss in energy arriving at the end of the line; there would be a slight drop in delivered current since the voltage is fixed by the generator. If you separate the two leads from a telephone and lay the pick-up loop (attached to a good set of headphones) next to one of the wires, you should be able to listen in on the time-varying B-field of the conversation.

22.13   Yes, a coil in a motor turning through a magnetic field will experience an induced emf and an induced current that will send energy back to the source, which is driving the motor. With no load, the motor produces (and returns to the power company) almost as much current as it draws, and so costs very little to run. A free-turning motor must have its speed limited by the back-current it generates — with no losses (in the bearings, etc.) the motor will speed up until the back-current equals the driving current and it can no longer accelerate. With a load, the motor does work and draws energy in excess of what it returns via the back-current. In real life, a motor also produces a good deal of thermal energy via friction and if it is to operate continuously for any long period of time, it must be cooled (usually by forced air). A jammed motor will not turn and not generate a back-current. The driving current (which depends on the motor's resistance) now undiminished by any back-current is too great. The wiring will heat up and the insulation will begin to burn off. If the process is not stopped soon, the motor will be destroyed.

22.15   Recall that the B-field outside a very long tight solenoid approaches zero. Each meter reads the potential drop across the resistor adjacent to it, the resistor with which it forms a closed circuit excluding the solenoid. The induced emf equals the difference between the two meter readings, or alternatively, the sum of the potential differences across the resistors in the central circuit (1-9-10-2-7-8-1). The seemingly strange thing here is that the "voltage" measured between 1 and 2 actually depends on how you hook up the meter. The varying flux provides energy to the induced current (energy coming from the power source driving current through the solenoid). In part (b) both meters would read $I_i R_2$.

Answers to Multiple Choice Questions

  1. e   3. c   5. a   7. a   9. c   11. c   13. b

15. d   17. d   19. b   21. b

Solutions to Problems

22.3 $\Phi_M = BA$, so $B = (6.0 \text{ mWb})/(0.005\,0 \text{ m}^2) = 1.2 \text{ Wb/m}^2$.

22.7   From Eq.(22.3), emf $= -(1)(0.25 \text{ m}^2)(-0.40 \text{ T})/(0.200 \text{ s}) = 0.50 \text{ V}$.

22.15   The bulb would not light just as a voltmeter across the tips would not indicate a voltage difference. The meter and its leads would have the same voltage induced across them and once attached to the wing no current would circulate. In other words, the flux through the closed loop of meter-leads-wings does not change if the field does not change. You could read a voltage difference if the meter stayed on the ground and still managed to remain connected to the plane.

22.21   From Eq. (22.3), $\mathscr{E}$ = -220(-20 mT/s)[$\pi(0.10)^2$] = 0.138 V. $CV = Q$ = 4.1 μC.

22.29 $\Delta\Phi_M = BA - (-BA) = 2BA$, so from the result in Problem 26 $B = QR/2NA$ = 0.83 mT.

22.35   The maximum voltage is 100 V. $\omega = 2\pi f$ = 376.99, so f = 60.000 Hz.

22.53   Across R, $V = IR$ = (1.25 mA)(20 Ω) = 25 mV = induced emf. Given the direction of motion of the charges in the disk, all inducted force and hence all current must be radial. Then the disk can be treated as a collection of wire segments radiating outward from the center and each in turn instantaneously in contact with the resistor at both ends. One such segment-resistor combination forms a loop the flux through which is changing at a rate proportional to the area being swept out by the segment per unit time. That rate is $\frac{1}{2}r^2\omega$, and so $\mathscr{E} = B\frac{1}{2}r^2\omega = B\frac{1}{2}r^2 2\pi f$. Then B = (25 mV)/(0.25 m)$^2\pi$(6 Hz) = 21 mT.

22.61   For plastic, $\mu \approx \mu_0$. From Eq. (22.9), then, $L/l \approx \mu_0 N^2 A/l^2 = \mu_0 A(N/l)^2$ = 6.32 mH/m; $PE_M = \frac{1}{2}LI^2$, and the energy per unit length, [$\frac{1}{2}I^2(L/l)$] is 79 mJ/m.

22.65   (a) $I_m = V/R$ = 2.0 A;   (b) L/R = 4.0 s;   (c) $\approx 5L/R \approx$ 20 s.

22.73   The self-induced emf is equal to the time rate of change of the flux linkage via Eq. (22.3); emf = -(210)(2 × 10 mWb)/(200 ms) = 21 V of a polarity which resists the change in current.

Answers to Discussion Questions

23.3   Unlike ac, dc does not induce currents in nearby
        conductors, thereby reducing the transmitted power. dc can
        be transmitted at higher voltages since it does not peak.
        The same *rms* voltage for an ac signal will rise to a peak
        voltage (1.414 times higher) and be more troublesome as
        far as insulation and breakdown of air is concerned).

23.7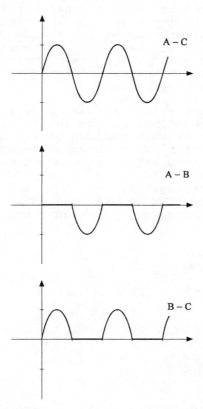

Across A-C, the scope reads the sinusoidal voltage induced
on the output side of the transformer. Across A-B, the
scope reads the voltage across the diode, which is zero
when current is flowing through (when the diode acts like
a closed switch) and equals the transformer output voltage
when no current is flowing (when the diode acts like an
open switch). Across B-C, the meter reads the voltage
across the resistor, which equals the transformer output
voltage when current is flowing and equals zero when no
current is flowing.

23.9   Put the tester across the terminals of the fuse holder without the fuse in place — if all is well, the side coming into the house will be hot, the other will be floating, there will be a sizeable potential difference and the neon will light. Now screw in the fuse and repeat the last test. If the fuse is good and if it is properly seated in the holder, there will be no appreciable voltage drop across it and the tester will not light. If it does, either the fuse is no good or it's not making good contact in the socket.

23.11   When $f \to 0$ (that is, dc), the ratio is 1 (whatever low-frequency signals were in the input are for the most part in the output). This arrangement passes low frequencies on to the next circuit. When $f \to \infty$, the ratio approaches 0; little or no high-frequency voltage appears across C. At $f \approx 0$, the capacitor is open circuited. At $f \approx \infty$, the capacitor is short circuited.

23.15   It's a one-transistor AM radio. The antenna feeds a tiny voltage (more or less, proportional to the length of the antenna wire) to the tuning circuit made up of the antenna coil and the variable capacitor. When tuned to resonate at the frequency of a particular station, current is maximum and the signal is passed on to the diode, which rectifies it (that is, it cuts off the negative portion). The transistor amplifies the positive signal so it will easily power the headphones, which respond to the envelope of the signal curve.

Answers to Multiple Choice Questions

1. b   3. d   5. e   7. c   9. b   11. e   13. a

15. b

Solutions to Problems

23.5   From Eq. (23.8), $V_m = 1.414 \, V$, since $V = 100 \, V$, $V_m = 141$ V.

23.15   From Eqs. (23.18) and (23.19), $V = IX_c = I/2\pi fC$; $C = I/2\pi fV$
= (1.00 A)/2π(60 Hz)(120 V) = 22 μF.

23.23   The voltmeter shows that the source is OK. If all
functioned properly R_e = 5.0 Ω + 10 Ω + 50 Ω = 65 Ω; I =
V/R =(120 V)/(65 Ω) = 1.8 A and the fuse wouldn't blow.
One of the resistors is malfunctioning, probably shorted
internally. Even if both the 10 Ω and the 5.0 Ω were
shorted the current would still be only 2.4 A, so it must
be the 50 Ω that's bad; (120 V)/(15 Ω) = 8.0 A.

23.49   From Eqs.(23.31) and (23.32), $Z = V/I$ = 2307.7 Ω =
$\sqrt{R^2 + X^2}$; then R = 1.7 kΩ.

23.55   From Eq.(23.35), $L = 1/4\pi^2 f_0^2 C = 1/4\pi^2 (40\ \text{Khz})^2 (300\ \text{pF})$ =
52.8 mH.

23.59 V_s = 30 V; from Eq.(23.38), $V_p I_p = V_s I_s$, so I_s = 4.0 A.

23.65 $Z = V/I$ = 100 Ω; $Z = \sqrt{R^2 + X^2}$; X = 97.98 Ω = $X_c = 1/2\pi fC$;
then C = 27 μF.

23.71   efficiency = output/input = (45 kW)/(45 kW + 300 W +
500 W) = 98.3%.

Answers to Discussion Questions

24.3 A hairdryer radiates radiowaves that can be picked up as noise in the picture on a nearby TV set.

24.7 The modern theory envisions light as both wave and particle; light energy is quantized but the propagation of those packets of energy is determined by its wave nature. Newton's version had a remarkably similar wave-particle structure: light was particulate but the particles were guided through space by wave patterns they set up in the aether.

24.9 The radiation is absorbed by the moist meat because it contains water. The plate is dry and stays cool. The water molecules in an ice cube cannot undergo rotational motion and will not absorb microwaves until some of the surface has melted.

24.13 A pulse can be a perfectly good wave. It's not obvious what is waving in a lightwave — it's certainly not a material aether. Later on, we'll talk about probability waves, but for the time being, let's say that the electromagnetic field itself waves.

24.15 The sphere is in equilibrium with its weight down, balanced by an upward force exerted by the beam. Radiant energy can therefore transfer momentum and exert pressure. A sail craft for space travel is quite possible.

Answers to Multiple Choice Questions

  1. c   3. b   5. a   7. b   9. a   11. d   13. c

15. d   17. b   19. e   21. c

Solutions to Problems

24.7 $E = (20 \text{ V/m}) \cos 0 = 20 \text{ V/m}$.

24.15 $f_0 = 1/(2\pi\sqrt{LC})$, so $L = 1/4\pi^2 f_0^2 C = 1/4\pi^2 (100 \times 10^6 \text{ Hz})^2 (0.5 \times 10^{-12} \text{ F}) = 5 \ \mu\text{H}$.

24.25 $k(x - vt) = kx - kvt$, but $k = 2\pi/\lambda$ and $v = f\lambda$, hence $kvt = (2\pi/\lambda)(f\lambda)t = 2\pi ft$. But $\omega = 2\pi f$, hence the phase is $(kx - \omega t)$.

24.29 From Eq. (24.6) $E = cB$, so $E_0 = cB_0 = 2.0 \times 10^2 \text{ V/m}$. Then $B_0 = 6.7 \times 10^{-7} \text{ T}$.

24.35 $E_{max} = 1.00 \times 10^4$ eV. Since by Eq.(24.9) $E = hf = hc/\lambda$,
        $\lambda = (4.136 \times 10^{-15}$ eV/Hz$)(3.00 \times 10^8$ m/s$)/(1.00 \times 10^4$ eV$)$
        $= 1.24 \times 10^{-10}$ m.

Answers to Discussion Questions

25.1 (a) It's mostly diffuse, although there is a little specular. (b) Flat paints have a diffuse, frequency-independent surface reflection that results in a white haze. (c) The index of water is between that of the fibers and the air so there is less diffuse white light reflected when it's wet. There will be a whitish haze over the dry painting.

25.7 The mirror is aimed too far toward her middle for her to be looking at herself. She's looking at you, which makes the picture strangely interesting.

25.9 The gentleman is standing where we are and her image is shifted too far to the right to be produced by a flat mirror parallel to the bar.

25.11 The resonance in the UV in part accounts for the effective absorption of UV by glass, which is essentially opaque to it. Only in great thicknesses will the weak absorption of red and blue produce a greenish tint.

25.13 (a) Red. (b) Since $C = W - R$, cyan ink "eats" red; hence, the ink will appear black. (c) It absorbs blue and produces more contrast between blue sky and white clouds.

25.17 (a) $C = B + G$ — the filter absorbs B; hence, only G emerges. (b) $Y = R + G$ — the filter passes $R + B$ and absorbs G; hence, R is transmitted. (c) $(B + G) + (R + B)$ $= (R + B + G) + B = W + B$, unsaturated blue. (d) $(R + G + B) - (R) - (G) = B$.

Answers to Multiple Choice Questions

1. c   3. d   5. e   7. a   9. d   11. d   13. d

15. d   17. c   19. e   21. d

Solutions to Problems

25.5 Since angles of incidence and reflection are measured with respect to the perpendicular to the reflecting surface, both are $\approx 90°$.

25.11 12.5 cm, since the observer-image distance will be twice that.

25.19   Draw a line from 0 perpendicular to the plane of the mirror and extend the lines $\overline{M_1 M_2}$ and $\overline{C_1' C_2'}$ to meet the perpendicular at points A and B respectively. Triangles OAM_2 and OBC_2' are similar and so are OAM_1 and OBC_1' hence $H/d = h/(d + s_0)$ and so $H = hd/(d + s_0)$.

25.25   From Eq. (25.2), $c/v = n = 2.42$; $v = 1.24 \times 10^8$ m/s.

25.33   Recall Fig. (25.3) and the accompanying discussion: $n_a/n_w = d_A/d_R = 1/1.333 = 0.750 = 3/4$.

25.39   From Eq. (25.5), $\sin \theta_c = (1.33/2.40)$; $\theta_c = 33.7°$.

25.49   Light entering at glancing incidence is transmitted at the critical angle and those rays limit the cone of light reaching the fish. $\sin \theta_c = 1/1.333$, so $\theta_c = 49°$ and the cone-angle is twice this or $98°$.

Answers to Discussion Questions

26.1   The focal length increases because the rays are not bent as strongly at the water-glass interface.

26.3   The focal length depends on the index of refraction and that depends on the wavelength.

26.5   To estimate the focal length of a converging lens, form a real image of a very distant object; the image-distance then approaches the focal length. Taking the positive and negative lenses, place the two in contact, shine in parallel light and measure f, knowing that $1/f = 1/f_+ + 1/f_-$, where f_+ is given and f_- is to be found.

26.9   The radius of curvative is ∞ and so is f. That means the object- and image-distances must have equal magnitudes. Thus, the magnification is +1.

26.13 The target is at one of the two foci of the hyperboloid and rays reflected from it appear to come from the other focus, $F_1(H)$, but this is also a focus, $F_1(E)$, of the ellipsoidal mirror. Rays appearing to come from one focus of the ellipsoid, after reflecting off it, converge to the other focus, $F_2(E)$, at the film plane.

26.15 The object has a diameter d, where $d(1200) = 0.0005$ m, so $d = 4.2 \times 10^{-7}$ m, which is the same as the wavelength of violet light. We cannot hope to see objects that are smaller than the probe being used; namely, the wavelength of light. The amount of diffraction will then obscure the image completely. So the details we wish to observe cannot be finer than λ, which puts a practical limit on the magnification. A 12 000X microscope will only make larger blurred images showing no more detail.

26.17 The filament is located at one focus of the ellipsoid and the reflected light is made to converge at the other focus, which also corresponds to the front focal point of the lens combination.

Answers to Multiple Choice Questions

   1. b   3. b   5. d   7. b   9. a   11. e   13. b

   15. d   17. a   19. c   21. e

Solutions to Problems

26.7   From Eq. (26.9), $1/s_o + 1/s_i = 1/f$, so $1/s_i = 1/(60.0 \text{ cm}) - 1/(100.0 \text{ cm}) = 0.006\,67$; $s_i = 150$ cm.

26.16   From 1/30 to 1/60 to 1/125 to 1/250 to 1/500 is 4 doublings of speed and halvings of energy. Hence we must go from $f/16$ to $f/11$ to $f/8$ to $f/5.6$ to $f/4$.

26.19   For the standard observer, from Eq. (26.14), $M_A = d_n/f = d_n\mathcal{D} = (0.254 \text{ m})\mathcal{D} = 8X$; then $\mathcal{D} = 31.5$ D and $f = 0.03$ m. The lens is 3 cm above the surface.

26.31   The total separation is 10.0 cm $= f_o + L + f_E = 1.0$ cm $+ L + 3.0$ cm; $L = 6.0$ cm. [See Fig. (26.29).]

26.39   From Eq. (26.9), $1/s_o + 1/s_i = 1/f$ and hence $f = s_o s_i/(s_o + s_i)$. We have $s_o = 0.10$ m and $s_i = 0.30$ m, so $f = 0.075$ m. When $s_o = 0.025$ m, $1/s_i = 1/f - 1/s_o = 13.33 \text{ m}^{-1} - 40.00 \text{ m}^{-1}$, so $s_i = -3.8$ cm. At first the image was real, inverted and magnified; after the jump it became virtual, erect, and magnified.

26.45 $M_T = y_i/y_0$, but since the two triangles are similar it follows that $M_T = -s_i/s_o$ [Eq. (26.12)] where the image is again inverted. The greater the s_i, the more the magnification.

26.53   The image will be inverted if it's to be real so the set must be upside down or else something more will be needed to flip the image. From Eq. (26.12), $M_T = -3 = -s_i/s_o$, so $s_i = 3s_o$ and from Eq. (26.9), $1/s_o + 1/3s_o = 1/(0.60 \text{ m})$. Then $s_o = 0.80$ m and hence $s_o + s_i = 0.80$ m $+ 3(0.80$ m$) = 3.2$ m.

26.61   (a) From Eqs. (26.23-25), $1/s_o + 1/s_i = 1/(0.30 \text{ m}) + 1/(9.0 \text{ m}) = 1/f = -2/R$; then $R = -0.58$ m. (b) $M_T = -s_i/s_o = (-9.0 \text{ m})/(0.30 \text{ m}) = -30$. Then the image is $(5.0 \text{ cm})(30) = 1.5$ m tall, real, and inverted.

26.65   From Eq. (26.24), $1/(500 \times 10^3 \text{ m}) + 1/s_i = -2/(-1.0 \text{ m})$; then $s_i = 0.50$ m. $M_T = -(0.50 \text{ m})/(5 \times 10^5 \text{ m}) = 10^{-6}$. So the image size is 2.0×10^{-6} m.

26.73   If s_i is the image-distance in the spherical mirror and β is the angle the image subtends, $\beta \approx y_{is}/(5.0 \text{ m} - s_i)$ where y_{is} is the image height and 5.0 m $- s_i$ is the distance from the observer to her image, s_i being negative (the approximation is good for small β). In the plane mirror $2\beta \approx (1.0 \text{ m})/(10 \text{ m})$, where 1.0 m is the subject height and the subject-image distance is 10 m. Then $\beta = 0.05$ rad. $M_T =$

$y_{is}/(1.0 \text{ m}) = -s_i/(5.0 \text{ m})$; substituting this for y_{is} into the expression for β we get $0.05 \text{ rad} = [-s_i/(5.0 \text{ m})]/(5.0 \text{ m} - s_i)$ and $s_i = -1.666 \text{ m}$. $1/(5.0 \text{ m}) + 1/(-1.666 \text{ m}) = 1/f$; $f = -2.5 \text{ m}$.

Answers to Discussion Questions

27.1   Yes. Monochromatic light can be imagined as the
superposition of many waves with different polarization
states but all of the same wavelength and all infinitely
long. There cannot be any random phase changes because
each component is infinitely long and constant in phase.
So all will simply combine to form a constant polarization
of some sort.

27.5   At the angle in question — the polarization angle — the
glare is perfectly polarized normal to the plane of
incidence. The polarizer then transmits none of the glare
at all. Since the index of refraction of benzene is around
1.5 (versus 1.33 for water), $\tan \theta_p$, and hence θ_p, will
increase when the water is replaced by benzene. The
reflection will again become visible.

27.9   The speckle effect arises from the interference of light
reflected from adjacent regions of the surface. The more
coherent the illuminating light, the more apparent the
phenomenon. With a laser the speckles almost fill the
space in front of the illuminated surface.

27.11   For two holes on a horizontal
line the pattern is as shown here.
With the two apertures close
together, their two Airy patterns
are seen as one, but with Young's
fringes cutting out portions of
the pattern.

27.13   Set the arms so that the optical path length for each is
the same and then move one mirror outward until the
contrast degrades so much that the fringes vanish; then l_c
= $2d$, where d is the distance the mirror was moved.

27.17   (a) 5; (b) 6; (c) 3; (d) 2; (e) 4; (f) 1; (g) 7.

Answers to Multiple Choice Questions

1. c   3. a   5. b   7. a   9. c   11. b   13. c

15. b   17. a

Solutions to Problems

27.7   The angle between the polarization of the light and the transmission axis of the polarizer is 60°. From Eq.(27.1), then, $I = I_1 \cos^2 \theta = (160 \text{ W/m}^2) \cos^2 60° = \frac{1}{4} 160 \text{ W/m}^2 = 40 \text{ W/m}^2$.

27.11   The light from the first polarizer is at 10° and has an irradiance of $I_i \cos^2 30° = 0.75 I_i$; this makes an angle of 60° with the next filter, hence $I_2 = (0.75I_i) \cos^2 60° = 0.19I_i$.

27.23   From Eq.(27.11) (with $m = 1$), $d = m\lambda_0/2n_f = (500 \text{ nm})/2(1.36) = 1.84 \times 10^{-7} \text{ m}$.

27.31   The center-line is a nodal line, i.e. the sound is a minimum along it because the sources are 180° out-of-phase. The first <u>maximum</u> will occur now when the path length difference is $\frac{1}{2}m\lambda$ (Eq.(27.4)), hence $a \sin \theta = \frac{1}{2}m\lambda$, $ay_m/s = \frac{1}{2}m\lambda$ and $y_m = \frac{1}{2}m\lambda s/a = \frac{1}{2}mvs/af$ (Eqs.(27.7, 27.8)). Hence for $m = 1$,
$y_1 = \frac{1}{2}(1)(346 \text{ m/s})(10.0 \text{ m})/(5.00 \text{ m})(1000 \text{ Hz}) = 0.346 \text{ m}$.

27.43   The two minima bounding the central maximum correspond to $m = \pm 1$ in Eq.(27.13), so $\sin \theta_1 = \lambda/D = (461.9 \times 10^{-9} \text{ m})/(1.0 \times 10^{-4} \text{ m}) = 461.9 \times 10^{-5}$ and $\theta_1 = 0.26° = 4.6 \times 10^{-3}$ rad. The central band is twice this or 0.53°.

27.47   From Eq.(27.9), $\Delta y = s\lambda/a$ and as the number of lines per centimeter doubles a is halved and so Δy, the space between orders, is doubled.

27.51   From Eq.(27.15), $\theta_a = 1.22\lambda/D = 1.22(550 \times 10^{-9} \text{ m})/(5.08 \text{ m}) = 1.32 \times 10^{-7} \text{ rad} = 7.57 \times 10^{-6}$ degrees; multiplying by (60 min per degree)(60 s per min) yields 2.72×10^{-2} seconds of arc.

27.63   From Eq.(27.8), $\sin 20.0° = 1(500 \times 10^{-9} \text{ m})/a$, so $a = 1.4619 \times 10^{-6} \text{ m}$; then $\sin 18.0° = 1\lambda/(1.4619 \times 10^{-6} \text{ m})$, and $\lambda = 451.75$ nm. Hence by Eq.(25.4), $n = (500 \text{ nm})/(451.75 \text{ nm}) = 1.11$.

Answers to Discussion Questions

28.1 (a) You would see yourself just behind where you are at
the instant you spin around, moving backward in space and
time until you plop down into the chair. The most distant
image would take the longest time to reach you. (b) To see
Lincoln, you need only fly away at a speed greater than c,
overtake the wavefront corresponding to the desired event,
pass it, and turn around. Similarly, you can rush off at v
$\approx 240c$, travel for a month or so, set up a telescope and
watch yourself being born (assuming the great event
occurred near a window with the shades up). (c) Of course
none of this is possible, since v must be less than c.

28.5 No. The scissors' contact point is more a mathematical
notion than a physical one. There is no transport of
energy from one point in space to another; nothing
actually goes outward from the pivot along the blades. The
situation is the same for the overlap region of the two
laser beams: it moves faster than c, but doesn't
communicate anything. Another "thing" that can move faster
than c is the tip of a long shadow.

28.7 The speed will remain constant at c, but there will be a
Doppler shift lowering the frequency of the light. Since E
$= hf$, a drop in frequency corresponds to a reduction in
total energy. Because $E = pc$, the momentum must decrease
as well. Photons have zero rest-energy and that doesn't
change.

28.9 An electron moving along the wire, say, to the right, sees
all the electrons in the first wire to be at rest and all
the positive ions to be moving left — the electron sees
the other wire move left. The observer electron in wire 2
sees a contraction of wire 1 and a resulting increase of
positive charge density while the negative charge
distribution is unchanged. The observer electron and all
its fellows are attracted to the increased positive charge
density — the wires attract.

28.11 As $v \to c$, $\gamma \to \infty$, and the total energy of the object
becomes infinite. We can conclude that it takes an
infinite amount of energy to bring a body up to light
speed and therefore no object with mass can attain a speed
of c.

28.15   Rewrite Eq. (28.7) as

$$\frac{v_{PO}}{c} = \frac{\left(\dfrac{v_{PO'}}{c}\right) + \left(\dfrac{v_{O'O}}{c}\right)}{1 + \left(\dfrac{v_{PO'}}{c}\right)\left(\dfrac{v_{O'O}}{c}\right)}$$

If $v_{PO'}/c = 1$ or $v_{O'O}/c = 1$, then $v_{PO}/c = 1$. When $v_{PO'}$ and $v_{O'O}$ are in the same direction, both + or both –, the denominator is > 1 and v_{PO} is less than its classical value, which keeps v_{PO} from exceeding c. When the signs of $v_{PO'}$ and $v_{O'O}$ are different and the motions are tending to cancel (as when someone runs to the rear of a moving train and stands still with respect to the platform), the denominator is < 1 and v_{PO} is larger than the classical value. That situation allows a light beam moving at $v_{PO'} = +c$ in a system moving at $v_{O'O} = -c$ to still be seen by someone at relative rest to be traveling at c.

28.17   We generalize the assumption and posit that <u>an object at one location in the Universe cannot affect another object that is a finite distance away instantaneously</u>. Then it follows that nothing can pass from one to another without a lapse of time and therefore <u>nothing can travel infinitely fast</u>. If there is an upper-limit to speed, and if it is measured to be different in different inertial systems, then it is possible to exceed the speed limit, and that contradicts the premise.

28.19   Envision an object interacting with another distant object via a stream of interaction particles traveling at a finite speed, call it c. If the first object now rushes ahead at a speed in excess of c, it can overtake the interaction particles and interact with itself, which is forbidden by the first postulate. Hence, c is the upper limit to speed.

Answers to Multiple Choice Questions

  1. d   3. c   5. b   7. a   9. e   11. c   13. a

15. d   17. c

Solutions to Problems

28.7   From Eq. (28.2), $\Delta t_M = \Delta t_S/\sqrt{1 - v^2/c^2} = (60.0\ s)/(0.099\ 9) = 601\ s$. The observer sees everything happening in slow motion and everybody laughs together or not at all.

28.15 $\gamma = 1/\sqrt{1-0.600^2} = 1.25$, and from Eq. (28.4), $L_S = \gamma L_M =$
1.25(400.0 m) = 500 m.

28.21   1.000 ft = 0.304 8 m, 1.000 yd = 3.000 ft = 0.914 4 m. From
Eq. (28.5), $L_M/L_S = \sqrt{1 - v^2/c^2} = 0.914\,4$, so $v^2/c^2 =$
$1 - 0.836\,1$, and $v = 0.404\,8c$.

28.27 $L_M = L_S/\gamma = (1.000\text{ m})/1.25 = 0.800$ m. The time the bar
takes to travel its apparent length is then given by
(0.800 m)/(0.600c) = 4.45 ns.

28.33   Let y be perpendicular to the line of flight, then if the
mast has a proper length of L, $L_y = L\sin 21.0° = L_y' =$
$0.358\,4\ L$. The proper x-component of L is $L_x = L\cos 21.0° =$
$0.933\,6\ L$, whereas $L_x' = 0.933\,6\ L/\gamma = 0.488\,8\ L$. $\tan \theta' =$
$(0.358\,4\ L)/(0.488\,8\ L)$, so $\theta' = 36.3°$.

28.39   By Eq. (28.12) $E = KE + E_0$, so KE = 1.0 MeV.

28.59   From Eq. (28.8), $p = \gamma mv = \gamma mc^2 v/c^2$, which by Eq. (28.13) and
the information supplied means that $\gamma v(938.3\text{ MeV})/c^2 =$

100.0 MeV/c. Then $\gamma v/c = 0.106\,57 = \beta\gamma = \beta/\sqrt{1 - \beta^2}$,
$\beta^2/(1 - \beta^2) = 0.011\,357$, $\beta^2 = 0.011\,357/1.011\,357$, and $\beta =$
$v/c = 0.106\,0$. So $v = 0.106\,0c$.

Answers to Discussion Questions

29.1   The NaCl disassociates into ions; namely, Na^+ and Cl^-. The chlorine carries its extra electron to the positive anode, where it gives it up and becomes neutral. The positive sodium ion, deficient by one electron (having given it to the chlorine), migrates to the cathode, where it picks up an electron and becomes neutral. The sodium atoms are not soluble and come out on the cathode as metallic sodium. The net result is that an electron has in effect traveled from cathode to anode, and that's the current. Had we started with a water solution, the water would have participated in the electrolysis and made things more complicated. (Sodium is also violently unstable in water.)

29.3   If it's finite in size, one might immediately ask, How is the charge distributed (that is, where on the electron is it)? If the electron has no parts, its charge has no parts either, and it becomes difficult to imagine any complicated distribution. Indeed if the charge were, say, spread over the surface of the electron (if it has a surface) what would keep the electron from blowing up under the Coulomb repulsion? Perhaps mass and charge are inseparable in the sense that whatever has mass has charge and vice versa.

29.7   The atoms in the first block are driven into oscillation in the zy-plane by the E-field, which is transverse to the x-axis. The beam traveling toward the second block is polarized. Atoms in the second block can only oscillate along the z-axis, and radiate in the xy-plane.

29.11   The wave number 109 677 cm^{-1} is the Series Limit of the Lyman Series just as 27 419 cm^{-1} is the Series Limit of the Balmer Series. All the levels correspond to the wave numbers of the successive Series Limits. It appears as if the atom, too, must have some sort of internal level structure such that a transition from one to another liberates light of a certain wavelength.

Answers to Multiple Choice Questions

1. d   3. b   5. b   7. c   9. b   11. a   13. b

15. c   17. a

Solutions to Problems

29.5   The net charge is (50.0 A)(5.00 min)(60 s/min) = 15 000 C, or 0.155 faraday; hence since the valence of sodium is 1, 0.155 moles or 3.55 g have been deposited.

29.13   From Eq.(29.3), $\sin \theta = m\lambda/2d$, so $\sin \theta = 0.148\,5$ and $\theta = 8.54°$.

29.23   63.55 g is 1 gram-mole and contains 6.022×10^{23} atoms. Such a mass has a volume of (63.55 g)/(8.96 g/cm^3) = 7.092 6 cm^3, hence if each atom occupied a cube its volume would be
(7.092 6 cm^3)/(6.022×10^{23} atoms) = $1.177\,8 \times 10^{-23}$ cm^3 with a side length of 2.28×10^{-10} m.

29.27 $e/m = E/B^2 r$, so $1.758\,8 \times 10^{11}$ C/kg = (20.0 kV/m)/$B^2 \times$ (15.00 cm); solving for B yields $B = 8.71 \times 10^{-4}$ T.

29.29   From Eq.(28.10) and setting mc^2 at 0.511 MeV, the electron rest-energy, KE = $\gamma mc^2 - mc^2$ = 1.00 MeV = (0.511 MeV) ($\gamma - 1$). Then $\gamma = 2.956\,9$, $\beta = 0.941\,1 = v/c$, and $v = 2.82 \times 10^8$ m/s.

29.35   From Eq.(29.4), $R = \dfrac{4(9.0 \times 10^9\,\text{N} \cdot \text{m}^2/\text{C}^2)(26)(1.60 \times 10^{-19}\,\text{C})^2}{(6.6 \times 10^{-27}\,\text{kg})(1.5 \times 10^7\,\text{m/s})^2}$ = 1.6×10^{-14} m.

Answers to Discussion Questions

30.3   Classically, the re-emitted X-rays should come off in a
        spherical wave and both detectors should have picked up
        the radiation simultaneously. Since they didn't, the
        photon picture is upheld.

30.5   Just increase the energy imparted to the electron by a
        factor of N: $Nhf = KE_{max} + \phi$. The maximum KE increases and
        so the stopping potential must increase, too, since $KE_{max} =$
        eV_s. The work function is determined by the metal and does
        not change unless the metal itself is altered. On the
        other hand, $Nhf = \phi_0$, and so the threshold frequency $f_0 =$
        ϕ/Nh must decrease.

30.7   The light-quantum must provide an energy of at least E_a.
        Hence, the photon must have a minimum frequency $f_0 = E_a/h$,
        very much as in the photoelectric effect.

30.9   The initial momentum as seen in the center-of-mass system
        is zero, as is the final momentum. The electron is seen to
        be at rest after the collision. The initial energy is hf
        for the photon and $E = \gamma mc^2$ for the moving electron. The
        final energy is $E_0 = mc^2$ since the photon is gone and the
        electron is at rest (KE = 0). Conservation of Energy
        yields $hf + \gamma mc^2 = mc^2$, which implies $m > \gamma m$.

INITIAL   FINAL

$p = h/\lambda$ $v_f = 0$

$p = mv$

30.11   Atoms are pumped up to the 4th level from the ground
        state, emptying the latter. They immediately drop to the
        3rd level, which is metastable and thus unlikely to
        experience many spontaneous emissions. There is then a
        population inversion between the 3rd and the 2nd levels
        and a laser transition occurs from the former down to the
        latter. Rapid decay from the 2nd to the ground state
        ensures that the inversion will be maintained, an
        improvement over the 3-level system.

30.13   Provided the atom was being illuminated by the proper
        light, it was absorbing and re-emitting photons almost
        continuously as it bounced up and down, into and from, the
        lower-excited state. The key was that the higher level was
        metastable. An atom can only be in one excited state at a
        time and so whenever the atom was kicked into the higher

state, it could not absorb and re-emit via the lower one. The bright light from the atom blinked off whenever it went into the metastable state, and it blinked back on as soon as the atom dropped out of the metastable state.

Answers to Multiple Choice Questions

1. c   3. a   5. b   7. b   9. c   11. d   13. b

15. c   17. b

Solutions to Problems

30.3   From Eq. (30.4), $\lambda_{max}T = 0.002\ 898$ m·K. 33° C = 306 K, and at $T = 306$ K, $\lambda_{max} = 9.4\ \mu m$, i.e., infrared.

30.15   Using Table 30.2, $KE_{max} = hf - \phi = hc/(200\ nm) - 4.31 = 1.89$ eV.

30.21   By conservation of *kinetic* energy, 60 keV - 25 keV = 35 keV. The angle of scatter is irrelevant.

30.25   The energy difference is 3.03×10^{-19} J, and since $E_i - E_f = hf$ (Eq. (30.14)), $f = 4.57 \times 10^{14}$ Hz. $\lambda = c/f = 656$ nm.

30.33 $KE_{max} = eV_S = (1.60 \times 10^{-19}$ C$)(1.250$ V$) = 2.00 \times 10^{-19}$ J. And since $KE = \frac{1}{2}mv^2$, $v = \sqrt{2\ (KE_{max})/m_e} = 0.663 \times 10^6$ m/s.

30.37 $hf = \Delta(KE) = \frac{1}{2}m_e(v_i^2 - v_f^2) = 3.416 \times 10^{-15}$ J; dividing by h, $f = 5.16 \times 10^{18}$ Hz. $\lambda = c/f = 5.82 \times 10^{-11}$ m.

30.45   The period of the electron's orbit is $T = 2\pi r/v$, so $1/T = f = v/2\pi r$. From Eq. (30.15) $v_n = nh/2\pi m_e r_n$, so $f_n = nh/(2\pi r_n)^2 m_e$. Using Eq. (30.17) to substitute in for r_n, $f_n = m_e k^2 Z^2 e^4 / 2\pi n^3 \hbar^3$.

Answers to Discussion Questions

31.1 Since from Eq.(31.2) $\lambda = 1.226/\sqrt{V}$ nm (where V is the accelerating potential and also, for individual electrons, the kinetic energy in electron volts), the wavelength can be quite small. For example, at 100 kV, $\lambda = 0.004$ nm. Although electrons can be focused easily, X-rays of comparable wavelength are very difficult to focus (though progress is being made in that endeavor). Electrons (in vacuum to prevent scattering) pass through a thin slice of the sample where they are scattered off in a pattern that corresponds to the information. This beam is collected by the objective, which forms a magnified intermediate image. The electrons from that image are collected and the image further enlarged by the projection lens. The product of these two magnifications can be an enlargement of 200 000 or so.

31.5 According to classical theory, a particle passes through one hole, a wave through both holes. If either coil records the presence of an electron, we are dealing with the particlelike manifestation of the electron and the wavelike behavior will vanish. Thus, the theory maintains that the interference pattern will vanish as soon as either coil picks up the transit of an electron. With one coil, presumably, the induced current will produce an induced magnetic field, which will destroy the interference.

31.7 If the particle is confined, from Eq.(31.3) there will be a finite Δx and therefore a nonzero Δp, which is the minimum p the particle must have and which corresponds to a minimum KE $= p^2/2m$; this is the zero-point energy. The atoms in a box at absolute zero must themselves still be moving around with a zero-point energy.

31.9 A range of frequencies and wavelengths is needed to synthesize the packet and, hence, a range of frequency (energy) and wavelength (momentum) is always present, thus constituting an uncertainty in E and p for the particle. To make $\Delta p = 0$, we must use a monochromatic wave, one wavelength, one p, and no uncertainty. But where will the particle be if the wave is a perfect sine wave? Anywhere, and so $\Delta x = \infty$. If the packet shrinks so $\Delta x \to 0$, the number of sine waves needed to make the packet increases to infinity, and $\Delta p \to \infty$.

31.11 The peaks occur at the Alkali Metals, which have a single outer electron in an unfilled shell. This electron is shielded from the nucleus by all the filled shells below it, and so is held weakly. If the nuclear charge is +Ze, the electron "sees" a charge of just +e. Accordingly, the electron cloud is relatively large. Similarly, boron, aluminum, and gallium atoms have three outer electrons that see a nuclear charge of +3e and are strongly bound in small orbits. Thus, the curve rises and falls.

Answers to Multiple Choice Questions

1. b   3. d   5. a   7. d   9. d   11. c   13. d

15. a

Solutions to Problems

31.3   From Eq.(31.1), $\lambda = h/mv = 10h/m_ec = 2.4 \times 10^{-11}$ m.

31.7   The kinetic energy of the electrons in electron-volts is given by $eV = \frac{1}{2}mv^2 = \frac{1}{2}m(p/m)^2 = \frac{1}{2}p^2/m = \frac{1}{2}(h/\lambda)^2/m$; dividing by e, $V = \frac{1}{2}(h/\lambda)^2/em = 0.15$ kV.

31.11   The number of electrons that can occupy the shell corresponding to quantum number n can be calculated by counting the possible combinations of l, m_l, and m_s. The general formula is $2[1 + 3 + ... + (2n - 1)]$. Then K has 2; L has 8; M has 18; N has 32.

31.19   Taking the mean lifetime as Δt and applying Eq.(31.5), $\Delta t = 4.4 \times 10^{-24}$ s, and $\Delta E = \frac{1}{2}\hbar/\Delta t = (3.29 \times 10^{-16}$ eV·s$)/$ $(4.4 \times 10^{-24}$ s$) = 75$ MeV. Then $\Delta E/E = 9.8\%$.

31.35 $\Delta p/p = 1/1000$, so $\Delta p = 10^{-3}p = 10^{-3}mv$. From Eq.(31.4), $\Delta x \geq \frac{1}{2}\hbar/\Delta p = \frac{1}{2}\hbar/(10^{-3}mv) = 2.64 \times 10^{-29}$ m. The uncertainty in position, then, must be at least this great.

Answers to Discussion Questions

32.3   The atom-bomb trigger generates temperatures in excess of
        100 million K. The lithium is blasted by neutrons and
        converted into tritium, which then combines via fusion
        with deuterium, liberating large amounts of energy. The
        fast neutrons (≈ 14.1 MeV) then fission the surrounding U-
        238 "blanket" which can be as large as can be delivered
        since there's no concern about critical mass.
        As for producing tritium, it can be separated from sea
        water, but is usually produced by putting lithium-6 in a
        fission reactor.

32.5   The surface of the pellet is vaporized, driving the
        remainder violently inward. The inner core compresses to
        tremendous densities (1000 times that of water). As the
        temperature rises to over 100 million degrees, the
        thermonuclear processes begin and there is a mini-H-bomb
        explosion (about the equivalent of 50 lb of high
        explosives). A constant drop and blast is the ultimate
        goal. The formula is $^3_1\text{H} + {}^2_1\text{H} \rightarrow {}^4_2\text{He} + {}^1_0\text{n} + 17.6$ MeV.

32.7   The half-life is the time it takes the sample to decay to
        half its original value. The nuclear mean lifetime is
        $1/0.693 = 1.44$ times longer than the half-life — the
        average radionuclide will live that long. The half-life of
        an American human is 68 years; half the people born 68
        years before will be dead. But the mean human life is
        certainly not $1.44(68 \text{ y}) = 98$. The difference is that
        nuclei don't age, and we do. Our chances of surviving two
        half-lives (136 y) are essentially zero. By comparison,
        some few nuclei will go on for 100 half-lives and more.

Answers to Multiple Choice Questions

   1. d   3. b   5. b   7. d   9. b   11. a   13. d

15. b   17. a   19. e

Solutions to Problems

32.1   111: 50 protons, 111 - 50 = 61 neutrons.

32.11 $(1.542\ 748 \times 10^{-25}$ kg$)/(1.660\ 540 \times 10^{-27}$ kg/u$) = 92.906\ 38$ u,
        which matches the value given in the atomic table. If
        there were other stable isotopes, the atomic table value
        would be some average of their different weights and would
        differ from the value calculated here (see Problem 23).

32.15   From Eq.(32.1), R = 3.6 fm = (1.2 fm)$A^{1/3}$; A = 3^3 = 27, so the element is aluminum.

32.23   Suppose there are no other long-lived isotopes. Then the relative abundance of Br-81 is 49.3% and the atomic mass listed in the table can be expected to be 50.7%(78.918 336 u) + 49.3%(80.916 289 u) = 79.909. Since this is the exact figure given, there are almost certainly no other stable or very long lived isotopes.

32.29 Δm = 26 m_p + 28 m_n - m_{Fe-54} = 26(1.007 825 u) + 28(1.008 665 u) - 53.939 613 u = 0.506 457 u. This binding energy is associated with A nucleons, so E_B/A = $\Delta mc^2/A$ = 8.736 MeV.

32.33   (mass initially) - (mass finally) = (232.037 13 u) - [(228.028 73 u) + (4.002 603 u)] = 0.005 80 u. Multiplying by the conversion factor of 931.494 MeV/u yields 5.40 MeV.

32.39   (mass before) - (mass after) = Q; 8.023 828 u - 8.005 206 u = 0.018 622 u, which converted yields 17.346 MeV.

32.53   (9.012 182 u) + (4.002 603 u) - (12.000 000 u) - (1.008 665 u) = 0.006 120 u, which converted yields 5.700 74 MeV.

32.57   We need R, which by Eq.(32.13) equals $N\lambda$ where λ is the decay constant. To find N, we convert mass to number of atoms using 222 as the atomic mass: 1 mg → (10^{-6} kg)/ (222 kg/kmole) = 4.505 × 10^{-9} kmole → 2.713 × 10^{18} atoms = N. Then R = 5.7 × 10^{12} Bq.

Answers to Discussion Questions

33.1   The collider smashes a beam of electrons head-on into a
        beam of positrons with a combined energy of ≈ 100 GeV. This
        quantity is more than enough to create Z° bosons — this is
        a so-called Z° factory, albeit a rather feeble one. The
        two electron pulses are accelerated (up to 1 GeV),
        condensed, and injected into the linac (linear
        accelerator). They are joined by a previously processed
        positron bunch, which is further accelerated, along with
        the leading electron bunch, to the end of the linac. The
        trailing electron bunch is diverted to the side, where it
        produces a shower of positrons that is sent back to the
        start for processing so that it can join the next pulse of
        electrons. Meanwhile, the two high-energy bunches are
        turned around by magnets at the end of the linac and they
        collide in the detector.

        The drift-tube linac is a succession of tubular conductors
        that have voltages applied to them just at the right
        moment so the electrons (or positrons) are accelerated
        across each gap. The particles see no E-field inside the
        conductors and drift along them. As they move faster, the
        cylinders are longer, so the generator frequency can be
        constant.

33.3   For the most part, the basis for the distinction between
        matter and antimatter arises by comparison to the stuff of
        our ordinary existence. If a baryon (for example, Λ)
        decays to ordinary matter (Λ decays into a neutron), we
        call it matter. If it decays to antimatter ($\bar{\Lambda}$ decays into
        an antineutron), we call it antimatter. But the mesons are
        half-and-half (quark-antiquark), so we can expect
        ambiguity. There is no way to determine, and no need to
        determine, which pion is the antipion.

33.7   A negatively charged pion entered the chamber leaving a
        clear slightly curved track, which means that it had a
        good deal of linear momentum ($R \propto p$). It struck a proton
        and created two nonionizing neutral particles (K° and Λ°)
        that left no tracks. Being unstable, they decayed via the
        weak force (thus living long enough to traverse a
        substantial distance). The process can be represented as
        $\pi^- + p \to \Lambda^0 + K^0 \to \pi^- + p + \pi^- + \pi^+$ (see also Questions 5
        and 6).

Answers to Multiple Choice Questions

1. d   3. e   5. c   7. c   9. a   11. a   13. d

15. a   17. c   19. b

Solutions to Problems

33.7   No; the electron-lepton number is not conserved.

33.11   No, it is not forbidden. The reaction conserves charge, lepton number, and spin. And with a weak decay we needn't worry about strangeness.

33.15   Referring to Table 33.1 and subtracting the energy equivalents of final mass from initial mass yields (1115.6 MeV) - (139.6 MeV) - (938.3 MeV) = 37.7 MeV. The reaction, which proceeds by the weak force, needn't conserve strangeness.

33.27   Subtracting final energies from initial ones, (139.6 MeV) + (938.3 MeV) - (497.7 MeV) - (1115.6 MeV) = -535.4 MeV. Since the net energy is negative, the reaction cannot take place without a considerable amount of energy present as KE initially.

33.31   As a meson it's $q\bar{q}$; it must have charm so one quark is c; that gives a charge of +2/3 so the next quark must have $Q = -2/3$, $S = 0$, $C = 0$ and that makes it $\bar{u}$. So $D^0 = c\bar{u}$.

33.35 $d\bar{s} \rightarrow u\bar{d} + d\bar{u}$. The $\bar{s}$ decays via the weak force into a $\bar{d}$ plus energy which creates a u-$\bar{u}$ pair, yielding the original d plus $\bar{u}$ plus $u\bar{d}$.

33.39   Using Yukawa's result, $R \approx h/4\pi mc \approx hc/4\pi mc^2$. Then $mc^2 \approx hc/4\pi R \approx (1.239\ 8 \times 10^{-6}\ \text{eV·m})/4\pi(10^{-31}\ \text{m}) \approx 10 \times 10^{23}\ \text{eV} \approx 10^{15}\ \text{GeV}$. So $m = 10^{15}\ \text{GeV}/c^2$.